Biomass Processing Technologies

Biomass Processing Technologies

Editor

Deepali Mudia

scitus
academics

Biomass Processing Technologies

Edited by **Deepali Mudia**

Printed in 2017

ISBN: 978-1-68117-031-2

Library of Congress Control Number: 2015931826

SCITUS Academics LLC,
616, Corporate Way, Suite 2, 4766,
Valley Cottage, NY 10989

www.scitusacademics.com

Contents

Preface

Biomass is biological material derived from living, or recently living organisms. In the context of biomass for energy this is often used to mean plant based material, but biomass can equally apply to both animal and vegetable derived material.

Biomass processing technologies capture value from biomass in different ways because biomass resources have varying ratios of distinctly different molecular structures (essentially water soluble carbohydrates, water insoluble carbohydrates and hydrocarbons). Value adding technology must be tailored to the biomass type to achieve optimum outcomes. Lignocellulosic material (woody biomass) is overwhelmingly the most common form of biomass in agricultural, industrial, municipal, forest and natural environments. Tree based lignocellulosic biomass can be essentially harvested on demand therefore offers the most stable and scalable primary resource for biomass technologies. The abundance of lignocellulosic biomasses provides strategic advantages to the thermal technologies and to emerging cellulosic fermentation. There are six generic technologies for energy conversion based on: direct combustion for power; anaerobic digestion for methane rich gas; fermentation of sugars for alcohols; oil exaction for biodiesel; pyrolysis for biochar, gas and oils; and gasification for syngas. These can then be followed by an array of secondary processing options depending on specific final products. Anaerobic digestion, fermentation and oil extraction are suited to specific biomasses that have easily extractable oils and sugars or high water contents. Such biomass resources are mainly associated with the food chain and are strategically constrained. Digestion, oil extraction and fermentation of sugars are all well established commercially. Thermal technologies such as direct combustion, pyrolysis and gasification, can effectively process all forms of biomass, including complete utilisation of lignocellulosic materials. Combustion and gasification of biomass for heat and power is commercially proven. Of the thermal processing options, pyrolysis provides the most flexible product platform, being able to generate solid, liquid and gas outputs roughly analogous to their coal, oil and gas fossilised counterparts. However, there are still relatively few commercial pyrolysis operations using biomass feedstocks.

This book, provides a concise overview of all currently available biomass processing technologies, is a reference tool for students, professionals and young researchers.

Editor

Chapter 1

The Challenge of Bioenergies: An Overview

Nicolas Carels

Fundação Oswaldo Cruz (FIOCRUZ), Instituto Oswaldo Cruz (IOC), Laboratório de Genômica Funcional e Bioinformática, Rio de Janeiro, Brazil

INTRODUCTION

The rapid rise in the price of crude oil, the decrease in oil reserves, security concerns and greater recognition of the environmental impacts of fossil fuels have generated considerable interest in biofuels as an alternative energy source. The revolution in transportation that occurred at the beginning of the last century created dependence of Western economies on the combustion of hydrocarbon fuels. The invention of the electric light bulb by T.A. Edison led to the installation of the first energy distribution plant in 1883 (in Roselle, New Jersey) and subsequently the electric grid came to the world. Because of the high rate (typically >90%) and ease of alternative to constant voltage conversion or vice versa, as well as mechanical conversion of electricity, the electric grid became the nervous system of our civilization.

During the 20th century, humanity created a foundation in which electric applications are used in all segments of society. This revolution is now creating a system of pervasive computing and is preparing society for the era of nanotechnologies and robotics. However, the electric grid is dangerously dependent on the availability of carbon-based resources such as coal and natural gas (Song, 2006). Human civilization is now challenged with finding renewable and environmentally friendly energy sources for feeding both our electric grid and our economic growth. For the next two decades, fossil fuels will continue to be the most cost-effective energy resource. However, despite their higher costs, alternative energies have begun to be seriously investigated. It is the purpose of the present work to provide an overview of this issue.

Existing fossil fuels are believed to have originated over the course of millions (M) of years (yrs) from biochemical and geochemical transformations of organic substances that were present on the earth's surface. The geological storage of carbon in the form of fossil fuels can be viewed as one alternative route to the reduction/oxidation cycle of carbon that is occurring at the earth's surface. Numerous studies have shown that coal is formed from biochemical degradation and geochemical maturation of higher-plant materials that were originally generated via photosynthesis. Crude oil also shows fingerprint molecules such as phytane that testify to its plant origin (Johnston et al., 2007).

Crude oil contains various hydrocarbons that range from light gases (e.g., C1-C5) to heavy residues (Fialkov et al., 2008). These hydrocarbons are separated via distillation into three main products, namely naphtha, middle distillate and a residual fraction. Naphtha (boiling range 90-190°C) is mainly used for motor gasoline (C3-C12) and comprises approximately 20% of the total crude oil. The middle distillate can be separated into two categories consisting of kerosene-range products (light-end) and diesel-range products (heavy-end). The light-end middle distillates (25-30% of total crude oil, boiling range 150-260°C) are used for the manufacture of solvents, kerosene (C8-C16), commercial and military jet fuels (C3- C10) and light diesel fuel (diesel fuel No. 1, C8-C22). The heavy-end middle distillates (25- 30% of total crude oil, boiling range 190-400°C) are processed to produce diesel fuel No. 2 (C10-C25) and heating oils (15-20%) (Kaplan et al., 1997). Lubricating oil (>C18) accounts for approximately 5-7% of the total crude. In addition to the alkane fraction, there is also a fraction of polycyclic aromatic hydrocarbons (PAH) whose relative importance in the composition of distillates increases from kerosene to heating oil. Beside the gas with green house effects (GHG) and their consequences on climate changes, it is the PAH that cause the most important environmental concerns because of oil spills. Actually, PAHs are rather resistant to microbial degradation (bioremediation) under anaerobic conditions, with the consequence that their removal from impacted environments is very slow. PAHs are also a source of concern regarding human health because they may stick to DNA, resulting in deleterious mutations and ultimately cancer.

World oil consumption is approximately 79 M barrels (bl) per day (a barrel is 159 l). The transport sector represents 50% of oil consumption (or 20% of energy consumption) and has an annual average growth rate of >2% per year. Energy demand worldwide is expected to rise by approximately 50%-60% over the next 20 years, reaching 112 M bl/day (Song, 2006). Most of the "cheap oil" that was (relatively) easily removed from the ground (and sold for US$ 20-30/bl) is already gone, and fossil fuel prices have risen over the past decades. The price of crude oil increased from US$ 18/bl in 1990 to US$ >100/bl in 2008 (Tan et al., 2008). Fuel prices will probably continue to increase in the future because of the following factors: (i) fossil fuel resources are limited; (ii) there is a lack of balance between supply and demand; (iii) the demand for fossil fuel is rising rapidly; and (iv) geopolitical instability and international conflicts are increasing (Rout et al., 2008).

With fossil fuel prices at US$ 45/bl, renewable energy from a range of biofuels is becoming economically competitive. Biodiesel fuel usually costs over US$ 0.5/l (as compared with US$ 0.35/l for fossil diesel fuel – see in Zhang et al., 2003) such that the increase in the use of biodiesel has been particularly rapid over the two last decades. It has grown from essentially zero in 1995 to more than 20 billion (G) liters (l) in 2009 (in the US as well as in Brazil). Brazil and the United

States (US) have the largest biofuel programs in the world, with the European Union (EU) ranking in the third position.

Biofuels can potentially mitigate greenhouse gases, can provide means of energy independence and may even provide new employment opportunities. Because they are compatible with existing technologies, they will also alleviate natural resistance to change and may act as a medium to allow a "smooth" transition to alternative technologies (Shinnar & Citro, 2006) and regulations. Many countries are now utilizing biofuel products resulting from agriculture and forestry. The most commonly cited advantages of biofuels are as follows: (i) they are available from common biomass sources; (ii) they are integrated into the combustion cycle of carbon dioxide (CO_2); (iii) they are compatible with environmental constraints; and (iv) they contribute to environmental sustainability. For example, biodiesel fuel prepared from vegetable oils or animal fats is compatible with common diesel engines and is therefore a potential alternative to fossil-based diesel fuel. Sulfate emissions resulting from biodiesel combustion are close to zero and the net contribution of CO_2 from biofuels when considering their entire life cycle (i.e., cultivation, oil production and conversion to biodiesel) can be low. The rate of pollutant emission over the life cycle of biofuels is comparable with that of fossil diesel fuel; however, because it is renewable, biofuel combustion does not result in CO_2 accumulation in the atmosphere (Agarwal, 2007). Actually, plants use solar energy to turn atmospheric CO_2 and water into organic carbon and hydrogen, thereby storing energy. The organic molecules are then broken down as the plant decays and the carbon is returned to the atmosphere as CO_2. When growing a crop for fuel, part of the biomass produced by the plant is used directly to produce energy and the CO_2 that was originally metabolized by the plant is returned to the atmosphere during the combustion process. This CO_2 is therefore"renewable" because it is simply a portion of the total CO_2 that is involved in the natural cycle. However, to produce a biofuel, a certain amount of energy is required. This sophisticated process of production needs a significant energy amount for growing, harvesting and processing the necessary biomass.

The concept of "well-to-wheef' (VIII'W) is used to characterize the energy consumption that is required to complete the entire process of production and transport of a fuel (Gnansounou eta!, 2009). The VIII'W analysis is often divided into the following five stages: (i) feedstock production, (ii) feedstock transport, (iii) fuel production, (iv) fuel distribution and (v) vehicle use. These stages can be divided even further into "well-to-tank" (VIll'T) and "tank-to-wheel" (TTW) processes. Fig. 3 in Agarwal (2007) shows that the rates of pollution with the corresponding increase in energy consumption of crude oil, natural gas, biomass, wind power are 1.3, 1.8, 7.1and 45 times lower, respectively, than that of coal. Nevertheless, biofuels (such as ethanol and biodiesel) fall on a line with a slope equal to crude oil. However, the intercept is lower, which demonstrates that the benefit of biofuels over fossil diesel fuel is due to the recycling of the former through photosynthesis. More advanced technologies (e.g., synthetic fuels based on biomass gasification or wind electricity) use virtually only renewable energy for the conversion processes and result in very low GHG emissions.

The CO_2 Balance

Carbon cycling does not occur with fossil fuel use because the released Cdoes not return to fossil resources.The amount of Cthat has accumulated in the atmosphere is very large and has been calculated to be ~780 billion (G) tons (t) as of 2002 (Song, 2006). CO_2 concentrations are measured by infrared absorption spectroscopy and, because of heterogeneity in the data, estimates may vary from the simple to the double (400 to 780 Gt). However, all measurements agree with a pattern of monotonic increase in CO_2. The concentration is now approximately 390 part per million (ppm), as compared to 310 ppm in 1955 (Song, 2006). Over the same period, the atmospheric temperature has increased by roughly 0.6-0.74°c and has resulted in numerous effects on the environment, climate (Odling-Smee, 2007; Rout eta!., 2008; Tomkiewicz, 2006), ecosystems (Yue eta!., 2010) and (consequently) human populations (Simoes eta!., 2010). For example, growing seasons are becoming longer with the consequence that the amount of carbon accumulating in terrestrial biomass increased by a factor of 7 between the 1980s and the 1990s. In addition to this terrestrial carbon sink (which plays an important role in slowing the rate of CO_2 increase), one must consider the oceanic carbon sink, which sequesters twice as much anthropogenic carbon as does the terrestrial sink (Ridgwell & Zeebe, 2005). Together, these sinks sequester approximately half of the ~24 Gt of anthropogenic carbon emitted each year. Although values of Cemissions should be considered carefully because of the heterogeneity among estimates, one may conclude that phototropic carbon sinks are keys for the mitigation of atmospheric Cconcentrations and, consequently, of global temperatures that will increase during the 21st century (Reay, 2007).

However, there is a limit to CO_2 mitigation by phototropic carbon sinks. With the rise in CO_2 concentrations in the atmosphere, falling oceanic pH levels will result in changes in ocean chemistry. This should decrease the efficiency of carbon recycling and, together with the decrease in precipitation, could induce a switch from oceans as a sink to a source of CO_2 by 2050. In fact, warmer soils will release more carbon and forests suffering increased drought will eventually decline. In addition, 40 Gt of Cthat is sequestered in 50 Gt of terrestrial biomass could be released because of changes in land use. Deforestation is the single biggest threat to the terrestrial carbon sink because woody biomass is the most efficient biological system for carbon sequestration (Miles & Kapos, 2008).

The world's forests cover ~4 billion (G) hectares (ha) or 30% of the earth's land surface. In 2005, 3.5 Gm^3 of wood were removed from 434 Gm^3 of forests; -60% of this amount was consumed for industrial purposes and the remainder was used for fuel. Terrestrial forests are either primary (36%) or modified (53%). The area of primary forest has been slowly decreasing at a rate of -6 Mha/ yr since the 1990s. This rate is especially high in Brazil and Indonesia; these two countries are responsible for the loss of 4.9 Mha of forests annually. Forest loss tends to occur in low-income countries (largely in the tropics), whereas higher income countries have reversed their trend of forest reduction (Kirilenko & Sedjo, 2007). Converting rainforests, peatlands, savannas, or grasslands to areas used for biofuel production and food crops in Brazil, Southeast Asia, and the United States creates a"carbon debt" by releasing 17 to 420 times more CO_2 than the annual greenhouse gas (GHG) reductions that these biofuels would provide by displacing fossil fuels. In contrast, biofuels from waste biomass or from biomass grown on marginal lands planted with perennial species incur little or no carbon debt (Fargione eta!., 2008;Searchinger eta!., 2008).Tropical deforestation is estimated to cause approximately one-quarter of anthropogenic carbon emissions, in addition to loss of biodiversity and environmental services. In Brazil, the Amazon forest has been cleared since the 1990s at a gross rate of ~25,000 km^2/yr (for comparison,

Belgium is 30,000 km^2). Kindermann et a!. (2008) estimated that a program providing a 10% reduction in deforestation

Currently, it is thought that the protection and enhancement of the terrestrial carbon sink is the best way to engineer climate against deleterious evolution (Reay, 2007). Worldwide monitoring of each country's contributions to GHG emissions should therefore be given the highest priority (Yang & Sirianni, 2010). In this spirit, the CarbonTracker is an integrated system for CO_2 measurement that assesses the net CO_2 exchange between the terrestrial biosphere and the atmosphere and that will be useful for the scientific community and for policymakers (Peters et a!., 2007). However, to establish effective GHG-reduction, without which CO_2 emissions will be soon unsustainable, governments need consistency and harmonization of policies. For example, it is the objective of the United Nations and other international organizations to raise the GDP/capita in developing countries to the same level as has been achieved in developed countries (e.g., US$ 9,000/capita). However, such politics would result in an increase in yearly global CO_2 emissions by a factor of 5 (Tomkiewicz, 2006) for the following reasons: (i) below the saturation level of US$ 25, 000/capita, there is a positive correlation between energy consumption per capita and purchasing power and (ii) the world average GDP/capita is US$ 5,210 (Tomkiewicz, 2006).

Increase in CO_2 emissions will therefore have to be compensated for by one or more of the following: (i) an increase in the global area planted with forests; (ii) a change in the technological paradigm (such as a commitment to convert to solar energy); or (iii) a decrease in human population size. Actually, global warming because of the increased use of fossil fuels matches predator-prey equations (Lonngren & Bai, 2008). The solutions to these equations are periodic in time. If humans continue to adhere to the current trends of fossil fuel consumption (as we have done since the beginning of the industrial revolution), we should observe large fluctuations in human population size in the future. Because earth and fossil fuel reserves are not infinite, the global-warming scenario leads (on a long timescale) to an eventual decrease in both the human population and the fossil fuel reserves, with the ultimate possibility being human extinction. Among catastrophic scenarios that could affect the human population because of climate change, we cite the prediction that one billion people could lack drinking water by 2050 if CO_2 emissions are not drastically reduced (Parry et al., 2008).

Greenhouse Gases and Their Management

The perturbation in atmospheric trace gases (e.g., SO_2, O_3, CO, CO_2, CH_4, NO_2, and CFCs, among others) is an important factor affecting climate change (Hopkin, 2007). In turn, climate change may promote changes in agricultural conditions that could have deleterious socioeconomic effects (Howden et al., 2007). Atmospheric trace gases have strong absorption bands in the infrared (IR) and interact with IR radiation emitted both by the earth's surface and the atmosphere. This directly influences the thermal structure of the atmospheric environment and contributes to the greenhouse effect. Gases such as NH3, SO_2 and their derivatives have lifetimes of only a few days, but they can have significant effects on the atmosphere (Begum, 2005). Emissions of N_2O and CH_4 are currently the dominant contributors. Although CO_2 is the main greenhouse gas in terms of volume, others must also be considered. In agricultural practices, the main culprit is nitrous oxide (N_2O), significant quantities of which are released from cultivated fields (particularly with the intensive use of fertilizers) (Snyder et al., 2009;Ceschia et al., 2010;Mander et al., 2010). Because N_2O is >300 times more potent as a GHG than is CO_2, even modest volumes can have significant impacts on the overall balance (Cherubini, 2010).

Harnessing the carbon sequestration capabilities of the terrestrial biosphere has been recognized as a potentially powerful, yet relatively low-cost, tool to offset carbon emissions (Dorian et al., 2006) and models for that purpose have been investigated (Werner et al., 2010). However, terrestrial carbon sequestration has been considered insufficient for meeting more than 25% of the CO_2 emissions reductions that are globally required by 2050. Given that carbon sinks are the best currently available scenario, an emissions credit system has been established to provide CO_2 emitters (companies or countries) with a means to satisfy the carbon liability associated with their release of carbon into the atmosphere. The emitter temporarily satisfies his liability by "storing" (for a fee) the equivalent carbon in a terrestrial carbon sink (such as a forest) (Sedjo & Marland, 2003). This concept is the application of the "willing-to-pay" principle within the international economic market. More simply, the right to emit CO_2 (in the form of a carbon credit) is compensated for by growing biomass that will sequester an equivalent amount of carbon. The marketing of carbon credits has been organized to allow for rewarding activities that result in the "permanent" immobilization of CO_2 in a nongaseous form. Ultimately, a carbon fee has been proposed that would be paid by industrial countries (in proportion to their emission contributions to GHG) to developing countries; these countries could then invest them in carbon mitigation practices (such as establishing or maintaining forest sinks) (Jones, 2010). The Kyoto Protocol is now legitimating activities such as revegetation, forest management, cropland management, grazing-land management, and also carbon sequestration in deep crustal layers (such as oil fields and deep saline aquifers) for trading with carbon credits (United Nations Framework Convention on Climate Change [UNFCCC], 2002). Principles of justice in proposals and policy approaches to avoided deforestation are also being pursued (Okereke & Dooley, 2010) through negotiations on Reducing Emissions from Deforestation and forest Degradation (REDD).

It has been estimated that the biological sink may attain a cumulative CO_2 sequestration of 100Gt over the next 50–100 years, with most of it in forest systems. This implies the capture of 10–20% from the anticipated net fossil fuel emissions until 2050 (Sedjo & Marland, 2003). However, carbon sequestered in the terrestrial biosphere may lack permanence. Forests may be harvested for timber that can be used to produce short-lived products or may be cleared for other purposes. Wild fires can release large amounts of sequestered carbon. Farmers may return to agricultural practices that release carbon that was previously captured. In that sense, terrestrial carbon sequestration may simply represent a delay in the flow of fossil fuel carbon to the atmosphere. However, economic incentives for carbon sequestration should increase permanent sequestration. That is, wherever and whenever there are incentives (payments) for carbon-sequestration services, one would expect more sequestration to occur than if no payments were made (Johnston & Holloway, 2007; Tollefson, 2008).

Carbon sequestration in living forests can be performed on lands with low productivity that are not suitable for agriculture or for intensive forestry and that are compatible with goals of biodiversity conservation over large areas. In contrast, to be economically viable, intensive crops for biofuels generally need land that is more productive. Intensive biofuel crops may compete with food production or even with the less-productive lands that are currently sheltering most of the earth's biodiversity (Huston & Marland, 2003; Miles & Kapos, 2008).

For example, this phenomena has been observed in Brazil, Indonesia and Malaysia, where cattle,soybean, sugarcane and palm oil may compete with standing forest (Darussalam, 2007; Laurance,2007Malhi et al.,2008; Stone, 2007; Venter et al., 2008). In Indonesia, this competition has disastrous consequences for wildlife. To resolve this problem, the Kyoto protocol and subsequent versions should include "wildlife credits" (in addition to carbon credits) to sustain wildlife and its buffering effect on human activities (Lovelock & Margulis,1974). This strategy would have the advantage of recognizing the fundamental roles played by ecosystem services and to begin to account for them (Mäler et al., 2008). New financial incentives are needed to act as a countervailing force to the economic pressures for deforestation (Jones, 2010). The recent agreement known as the "Bali Roadmap", which aims to extend the Kyoto Protocol beyond 2012, includes directives for providing compensation to rainforest-holding nations in exchange for control of deforestation and environment degradation. Such compensation could be managed either through international carbon markets or through voluntary funds. These directives have the potential to shift the balance of underlying economic market forces that currently favor deforestation by raising billions of dollars to pay for the ecosystem services provided by rainforests. However, to be effective they will require exceptional planning, execution and long-term follow-through. The new proposal also aims to reduce EU CO_2 emissions by 30% by 2020 if a global climate deal is reached in international negotiations (if not, the cut will be 20%) (Schiermeier,2008).The EU is also planning to protect its economy against carbon "dumpers" by applying leverage that aims to force companies that import goods from polluting countries to buy emissions permits (Barnet, 2008).

Typically, carbon-credit compensation funds are used in developing countries for establishing new long-term plantations (such as rubber trees or oil palm). One difficulty is that the actual goal of carbon sequestration can be negated in cases where the renter first illegally burns the original forest, earns the carbon-credit funds and subsequently establishes a new plantation that will never be as productive, in terms of carbon sequestration, as the original forest. In some regions, environmental crimes are not easily detected and may also not be "significantly" punished. Key recommendations to ensure the environmental sustainability of biofuels through certification (including international approaches and global monitoring) should help to control the process (Scarlat & Dallemand,2010).Despite these problems, the carbon-credit market was stabilized as of 2007 and is not expected to collapse any further (Haag, 2007). The next few years represent a unique opportunity (perhaps the last) to maintain the resilience of biodiversity and ecosystem services (Malhi et al., 2008; García-Montero et al., 2010; Hagerman et al., 2010).

Why Biofuels

Compared with an array of solar cells, plants are strikingly poor transducers of the sun's energy. Energy storage by photosynthesis is approximately <2 watts (W) per m^2. The important difference between plants and solar cells is that plants are very cheap. They are able to grow with a moderate supply of water, nutrients and CO_2 that they turn into stable organic compounds. No fuel technology is perfect, but the GHG crisis and concern over oil supplies means that diversifying the range of fuel options makes good sense; at the very least, such diversification places humanity on a healthy learning curve (Haig et al., 2011). Continuous increases in energy needs have encouraged governments to search for new alternatives to fossil fuels. The rationale is to facilitate the transition from a fossil-energy based economy to an economy based on renewable sources of energy. Numerous low emission scenarios have demonstrated that the goals of the Kyoto Protocol cannot be achieved without providing a large role for biofuels by 2050 in the global energy economy. Among the reasons why biofuels are appropriate for such a transition are the following: (i) their simplicity; (ii) their production via well-known agricultural technologies; (iii) their potential for mitigation of climate warming without complete restructuring of the current working energy system; (iv) the use of existing engines for their transportation (even considering the conventional turbofan used in aviation) (Kleiner,2007;Rothengatt er,2010); (v) their potential to facilitate worldwide mobilization around a common set of regulations; (vi) their potential as a directly available energy source with good public acceptance; (vii) their more uniform distribution than the distributions of fossil fuel and nuclear resources; and (viii) their potential to create benefits for rural areas, including employment creation.

What Are Biofuels

The term biofuel refers to liquid, gas and solid fuels that are predominantly produced from biomass. The production of biofuels may ignite concerns about security, the environment, trading, and socioeconomic issues related to the rural sector. Biofuels include bioethanol, biomethanol, vegetable oils, biodiesel, biogas, bio-synthetic gas (bio-syngas), bio-oil, biochar (charcoal created by biomass pyrolysis), Fischer-Tropsch liquids and biohydrogen. Biogas and bio-oil are primary products, the preliminary processing of which is almost reduced to collecting the raw material. Most traditional biofuels (such as ethanol from corn, wheat, or sugar beets and biodiesel from oil seeds) are produced from classic agricultural food crops that require high-quality agricultural land for growth. The biofuel economy will grow rapidly during the 21st century (Demirbas, 2008a).

Biogas

There are two basic procedures for transforming solid biomass into liquid or gaseous biofuels. The first is to transform it by microbiological fermentation (Gavrilescu & Chisti, 2005) (i.e., to convert the polysaccharides into alcohols such as bioethanol or biobutanol) or to convert raw plant biomass by anaerobic fermentation into biomethane (Demirbas & Balat, 2006). One of the main drawbacks of the anaerobic system is that it is slow (because of the small amount of energy that is available to the organisms). Therefore, the amount of methane that can be produced is limited and this technology is only sustainable under selected scenarios (Asam et al., 2011). However, the introduction of even a small installation for transforming agricultural and human wastes into methane can have an enormous effect on the living standards of small communities (Arthur et al., 2011; Parker, 2002).

The second procedure aims to thermochemically convert the total biomass into a synthesis gas of high calorific value (also called syngas, i.e., H_2+CO) with subsequent production of various liquid and gaseous fuels (Tijmensen et al., 2002). The production of syngas is a potential area for large-scale CO_2 conversion and utilization. The reforming of CO_2 to CH_4 has been extensively studied and reported on in the literature (Song & Pan, 2004). The catalytic reduction of CO_2 to form methanol (or even CH_4) using renewable energy sources could become a viable alternative to scarce or expensive fossil resources.

Biodiesel from plant oils and bioalcohol from sugar use only a portion of the total biomass. Next-generation processes are being developed to convert biomass to syngas (Baker & Keisler, 2011; Fagernäs et al., 2010) that can then be converted into fuels or chemicals by a synthetic process (the so-called Fischer-Tropsch, or FT, process).

$$CH0.8 + 0.7O_2\ CO + 0.4H_2O \quad (1)$$

$$CH0.8 + H_2O\ CO + 1.4H_2 \quad (2)$$

$$CO + H_2O\ CO_2 + H_2 \quad (3)$$

$$(2n+1)H_2 + nCO \rightarrow CnH_{2n+2} + nH_2O \quad (4)$$

Considering that coal inputs supply a 0.8 to 1 ratio of H/C, the whole FT process can be briefly written as follows. The partial oxidation of coal by oxygen gives equation (1). The interaction of water with carbon monoxide through "steam reforming" produces equation (2). Subsequently, the H/CO ratio is improved by "shifting" (transferring) the oxygen from the molecular water to CO, producing an additional hydrogen and carbon dioxide following equation (3). After removing the sulfur and carbon dioxide contaminants, the syngas is reacted over a catalyst in theFT reactor to produce high-quality dean fuels following the formula (4) (Greyvenstein et a!, 2008)

Biomass is more reactive than coal and (depending on the technology) is usually gasified at temperatures of between 550°C and 1,500°C and at pressures varying between 4 and 30 bars (Damartzis & Zabaniotou 2011 Leibold et a!, 2008; Steinberg, 2006).Typically, biomass is burned in an electrically heated furnace consisting of several multiple-tube units that can be heated separately up to 1,350°C Theis et a!,2006).Alternatively, the conversion of fossil fuel or biomass can be performed in hydrogen plasma. The temperature induced by an electric arc in hydrogen plasma is very high (~1,500°C); therefore, this technology produces hydrogen and CO gas with a conversion rate of near 100% (Steinberg, 2006). FT synthesis generates intermediate products for synthetic fuels (Liu eta!., 2007).The thermal efficiency of producing electricity and hydrogen through hydrogen plasma and carbon fuel cells varies from 87% to 92%, depending on the type of fuel and the biomass feedstock. This is more than twice as efficient as a conventional steam plant that burns coal and generates power with a ~38% efficiency.In addition, coupling hydrogen plasma and carbon fuel-cell technologies allows for a 75% reduction in CO_2 emission per unit of electricity (Steinberg, 2006).

Because FT produces predominantly linear hydrocarbon chains, this process is currently attracting considerable interest. FT has already been applied on a commercial scale by Sasol, Petro S.A. and Shell, mainly to produce transportation fuels and chemicals (the feedstock being coal or natural gas). This fuel option has several notable advantages. First, the FT process can produce hydrocarbons of different lengths (typically <C15, Liu eta! 2007) from any carbonaceous feedstocks; these hydrocarbons can then be refined to easily transportable liquid fuels. Secondly, because of their functional similarities to conventional refinery products, the synthetic fuels (synfuels) produced by theFT process (i) can be handled by existing transportation systems; (ii) can be stored in refueling infrastructure for petroleum products; (ii) are largely compatible with current vehicles; and (iv) can be blended with current petroleum fuels (Tijmensen et a!,2002). Thirdly, synfuels are of high quality (this is especially true for FT diesel), have a very high cetane number and are free of sulfur, nitrogen, aromatics, and other contaminants typically found in petroleum products. The principal drawbacks of the FT process are that the capital cost of FT-conversion plants is relatively high and that the energy efficiency for the production of FT liquids by conventional techniques is lower than the energy efficiency for the production of alternative fuels (Takeshita & Yamaji, 2008).

Bio-oil

Bio-oils are dark red-to-black liquids that are produced by biomass pyrolysis. Biomass is typically obtained from municipal wastes or from agricultural and forestry by-products (Demirbas, 2007). With an efficiency rate as high as -70%, pyrolysis is among the most efficient processes for biomass conversion. The density of the liquid is approximately 1,200 kg/ m3, which is higher than that of fuel oils and significantly higher than that of the original biomass. The gasification of bio-oil with pure oxygen and the further processing of syngas into synthetic fuel by theFT process, is being investigated; however, this process does not appear to be economically attractive (Demirbas & Balat, 2006).

Plant Oils

There has been interest in the use of virgin plant oils to fuel diesel engines. At least 2,000 oleaginous species, growing in almost all climates and latitudes, have been identified. Thereare more than 350 plant species that produce oil that could be used to power diesel engines (Goering et al., 1982). The plant oils are made up of 98% triglycerides and small amounts of mono- and diglycerides. There are basically two types of vegetable oils: those in which the majority of fatty acids are in C12 (e.g., palms) and those in which the majority of fatty acids are in C18. The direct use of plant oils (and/or blends of these oils with fossil fuels) has generally been considered to be unsatisfactory or impracticable for both direct and indirect diesel engines. Obvious problems include their high viscosity (Ramadhas et al., 2005), acidic composition, free fatty acid content, tendency to deposit carbon, tendency for lubricating-oil thickening, and gum formation because of oxidation polymerization during storage and combustion. When blending vegetable oils with fossil diesel fuel, the viscosity can be extensively adjusted. Based on EN 14214 recommendations, the maximum blending rate of most vegetable oils is B30 (30% plant oil/fossil diesel, v/v) (Abollé et al., 2008). The oil viscosity (because of the presence large triglycerides) can also be reduced by pyrolysis, which produces an alternative fuel for diesel engines (Lima et al., 2004). Using plant oils in blends also significantly increases their cloud points and thus limits their use to climatically compatible countries. 1.5.4 Bioalcohol Because of the energy crisis

Bioalcohol

Because of the energy crisis and climate warming, humanity faces the need for a huge short term supply of biofuels (see below).Bioethanol and biodiesel have been considered the best candidates for satisfying these needs and are what we consider the first generation of biofuels. Ethanol can be produced from a range of crops including sugarcane, sugar beets, maize, barley, potatoes, cassava, and mahua (Baker & Keisler 2011; Kremer & Fachetti 2000). Flexible-fuel motors have been developed that can burn hydrous ethanol/ gasoline blends in any combination, including pure ethanol. The automatic adjustment of combustion parameters is controlled electronically in these engines as a function of the oxygen level needed by the fuel in the tank (Marris, 2006). The so-called "gasohol" is a blend of ethanol and gasoline. Ethanol is produced via fermentation of a sugar slurry that is typically prepared from sugar or grain crops. The action of yeast on the sugar produces a solution that contains approximately 12% ethanol. The yeast invertase catalyzes the sucrose hydrolysis into glucose and fructose. Subsequently, yeast zymase converts the glucose and the fructose into ethanol.The alcohol can then be concentrated by distillation to produce up to 96% ethanol (hydrous ethanol). Ethanol is a polar solvent and its chemistry is very different from that of hydrocarbon fuels (which are non-polar solvents). As a result, blending ethanol into hydrocarbon fuels introduces some specific challenges.These challenges include (i) higher fuel volatility at low rates of ethanol/ gasoline blends, (ii) higher octane ratings, (iii) an increase in dissolved water content in motor gasoline that promotes heterogeneity of fuel blends and resulting engine corrosion and (iv) higher solvency. However, Akzo Nobel Surface Chemistry and the Lubrizol Corporation have developed and produced a low cost additive that makes it possible to blend ethanol with diesel fuel to obtain a stable and dear fuel (Lti et a!., 2004). This fuel is called "Dieshol". Biomethanol can be produced from biomass using bio-syngas obtained from the steam reforming processing of biomass. Biomethanol is considerably easier to recover from biomass than is bioethanol. However, sustainable methods of methanol production are not currently economically viable. The production of methanol from biomass is a cost-intensive chemical process. Therefore, under current conditions, only waste biomass, such as wood or municipal waste, is used to produce methanol.

Biodiesel

Biodiesel has the advantage that it can be used in any diesel engine without modification.It is produced by the transformation of renewable oils, such as those synthetized by plants, algae, bacteria and fungi. First-generation biodiesel is considered to be the result of a two stage process that involves (i) the crushing of raw material (typically oilseeds) in specialized mills to expel the oils and (ii) the transformation of oil into biodiesel. Free fatty acids (FFA) or triglycerides are converted into alkyl-esters by reaction with short-chain alcohols (such as methanol or ethanol) in the presence of a catalyst. The reaction involved in the conversion of FFA to alkyl-esters is called esterification, whereas that involved in the conversion of triglycerides is called transesterification. Fatty acid methyl-esters are only partly biological, as the methanol involved is generally produced from fossil methane (natural gas). However, biodiesel can also be produced by replacing methanol with ethanol, resulting in fatty acid ethyl-esters. If the ethanol is of biological origin, the product is fully biological. The purpose of the transesterification process is to lower the viscosity of the oil with transesterification being less expensive than the pyrolysis that is used in bio-oil processing. According to the EU standards for alternative diesel fuels, alkyl-esters in biodiesel must be ≥96.5 wt%. 1.5.6 The four generations of biofuels, the first generation of biofuels demonstrated that energy crops are technically feasible, but that no single solution exists to cover every situation (Venturi & Venturi, 2003). In addition, the production of first-generation biofuels is complicated by issues that are contrary to biofuel philosophy, such as the destruction of tropical rainforests (Kleiner, 2008). Tropical rainforests are the most efficient carbon sinks on earth. Therefore, if biofuels contribute to their destruction, this implies that the carbon balance of biofuels is negative. This consideration limits the viability of first-generation biofuels. It also comes with the corollary that raw materials for biofuel production will have to be diversified over the long term. Second-generation bioethanol is precisely an attempt to overcome this challenge. Second-generation bioethanol will be produced from lignocellulosic biomass, which is a more suitable source of renewable energy (Frondel & Peters, 2007; Tan et al., 2008; Tilman et al., 2007). Lignocellulose is obtained from inexpensive cellulosic biomass that is encountered throughout the world. However, the low-cost transformation of lignocellulose into bioethanol is still challenging. Some possible technologies involve genetic modification of plants, which is a source of concern for society. Whatever the future evolution of the technology, the introduction of energy policies is crucial to ensure that biomass ethanol is effectively developed to become a major source of renewable energy (Tan et al., 2008). Algae and cyanobacteria are far more efficient than higher plants in capturing solar energy and will surpass first- and second-generation biofuels in terms of energy capture per unit of surface area (Brennan & Owende, 2010). Algae are already used in pilot CO_2-sequestration units for emissions cleaning in some conventional power plants running on fossil fuels. This technique is called CO_2 filtration. Unfortunately, algae require capital for investing in reactors that can grow them, making CO_2 filtration an excellent opportunity for developing this technology. In that sense, algae can be regarded as a third-generation fuel. New methods and technologies for the production of second- (such as synfuels, Baker & Keisler, 2011) and www.intechopen. third-generation biodiesel fuels are being developed and will result in the modification of the definition of biofuels that is generally used in government regulations (Lois, 2007). Finally, one can also envision the exploration of photosynthetic mechanisms for biohydrogen and bioelectricity production. These would constitute fourth-generation biofuels (Gressel, 2008). The development of effective fourth-generation biofuels is not expected before the second half of the 21st century.

Plant Biofuels

Bioethanol

The technique of alcohol fermentation has been known for thousands of years. Ethanol distillation has been carried out for decades by industry because it has been part of the process of the regulation of sugar prices on the international market. Ethanol is regularly produced from the isomerose (high-fructose syrup) of grain crops such as maize or wheat and from sugar crops such as sugar beet or sugarcane. In Europe, sugar beet is preferred. This is especially true in countries such as the UK, France, Holland, Belgium and Germany, where it is highly productive, as 1 ha of this crop can produce 5.5 t of ethanol, (1 ha of wheat only produces 2.5 t of ethanol) (Demirbas & Balat, 2006). These numbers must be compared to the ethanol production from sugarcane, which reaches 7.5 t in Brazil (Bourne, 2007). The USA produces ethanol from corn, whereas India uses sugarcane, China uses sweet potatoes and Canada uses wheat. Countries such as China, Austria, Sweden, New Zealand, and even Ghana are now building their biofuels infrastructure around wood-based feedstocks (Herrera, 2006).

The growing area used for sugarcane production in Brazil accounts for 8 Mha (Brazil is 850 Mha). Sugarcane produces an eight-fold return on the energy that is used to produce it. One ton of sugarcane used for ethanol production represents a net economy in CO_2 emissions equivalent to 220.5 when compared with fossil fuel. Thus, if rain forest is not destroyed to grow the sugarcane, ethanol from Brazilian sugarcane reduces greenhouse gas emissions by the equivalent of 25.8 Mt CO_2/yr (Marris, 2006; Walter et al., 2010). Fortunately, the Amazon, the Pantanal and the Alto Rio Paraguai regions have been prohibited for sugarcane cultivation by government decree since 2009 to preserve these ecosystems. In 2009, ethanol accounted for approximately 47% of transport fuel used in Brazil. The "Flex" car fleet can use 100% of either ethanol or gasoline (Orellana & Neto, 2006). In fact, ethanol gives 20% to 30% fewer kilometers per liter than does gasoline and people adapt the blend in proportion to the best consumption/price ratio (Marris, 2006).

The ethanol export capacity of Brazil is currently ~8 Gl. The export-destination countries are mainly the US, the EU, Japan and Central America. Conservative estimates suggest that the area used for sugarcane production in Brazil should increase from 8 to 11 Mha by 2015. By government decree, the maximum possible area to be used for sugarcane cultivation has been limited to 64 Mha (i.e., 18.5% of national territory). In the short-to-medium term, Brazil is the only country that is able to sustain the emerging international ethanol market. For long-term establishment in the market, other countries, such as Australia, Columbia, Guatemala, India, Mexico and Thailand, will need to increase their exports (Orellana & Neto, 2006).

Brazil began ethanol production in 1973. At that time, it was heavily dependent on foreign crude oil, with nearly 80% of its oil being imported. It launched the program PROALCOOL in 1975 (Goldemberg et al., 2004) and began to offer subsidies and low-interest loans to bioethanol producers to increase existing capacity. A policy of price dumping was maintained by the government to boost the use of gasohol. The ethanol content of common gasoline was originally 5% and is now 25% by law (Pousa et al., 2007).

Bioethanol from Lignocellulosic Biomass

The most abundant sources of renewable carbon in the biosphere are plant structural polysaccharides. Approximately 1,011 t of these polymers (with an energy content equivalent to 640 Gt of oil) are synthesized annually (Proctor et al., 2005). For example, nonfood plant species for bioenergy production include Sorghum halepense, Arundo donax, Phalaris arundinacea (Raghu et al., 2006), poplar, switchgrass (Panicum virgatum), the hybrid grass Miscanthus x giganteus and big bluestem. These species are considered to have energetic, economic and environmental advantages over first-generation biofuel crops (Hill et al., 2006; Havlík et al., 2010). Switchgrass, for example, produces a net energy of 60 Giga Joule per hectare and per year (GJ/ha/yr) (Schmer et al., 2008). The potential terrestrial fuel yield from cellulosic biomass production (135 GJ/ha/yr) is somewhat higher than that from corn (85 GJ/ha/yr) or soybean biodiesel (18 GJ/ha/yr). The optimal types of specialized biofuel crops are likely to be perennial and indigenous species that are well adapted for growth on marginal lands.

In tropical and Mediterranean countries, eucalyptus is a fast-growing woody species that is cultivated for biomass production. In wet and temperate countries, high-yielding varieties of willow (Salix nigra), Miscanthus (a high-yielding rhizomatous grass that yields up to 26 t of dry matter/ha/yr) and poplar are available. These energy crops require relatively low chemical and energy inputs compared with conventional crop production and they are able to grow on marginal lands (thus avoiding the problem of competition with food crops).

Considering an Ireland-based scenario, the utilization of Miscanthus and willow for heat and electricity generation would allow for savings of as much as 5.2% of 2004 GHG emissions while using only 4.6% of the total agricultural area (Styles & Jones, 2008). It has been estimated that lignocellulosic biomass could contribute 70-100 exajoules (1 exajoule = 1,000,000,000 gigajoules) by 2020 (Gielen et al., 2002).

Poplar is a candidate for short rotations of ~5 years. Poplar disperses its seeds and pollen much farther than do other crops, it does so for many years before harvesting and it has many wild relatives with which it can outcross. In addition, poplar can be multiplied vegetatively, which would allow for the valorization of low-lignin transformants through the multiplication of sterile accessions. The biotechnology of poplar has been dominated for several years and its genome has been sequenced.

Trees not only can achieve a lignocellulosic energy-conversion factor of 16 (compared with 1–1.5 for corn and 8–10 for sugarcane), but they can also be grown on marginal lands, thus reducing competition with food crops. The world consumption of wood is 3.4 Gm^3/yr and will substantially increase with the production of ethanol from biomass. The development of high-yield plantations is essential to sustain the increased demand for wood (Fenning et al., 2008). Small towns, schools, buses, ski resorts and factories in Sweden and Austria have long relied on the byproducts of the forest industry to produce liquid and solid fuel (Herrera, 2006).

Biotechnology and systems biology can be envisaged for plant breeding. Many plant species used for bioenergy production are wild to semi-domesticated. Molecular approaches can speed up domestication and productivity (Chen & Dixon, 2007).

A number of candidate genes for domestication traits have been identified by comparing the genomes of poplar, rice and Arabidopsis for large-scale gene function and expression. The genes investigated were involved in synthesis of cellulose and hemicellulose, as well as in various morphological growth characteristics (such as height, branch number and stem thickness) (Ragauskas et al., 2006; Chapple et al., 2007; Sticklen, 2008). Transgenic plants that overexpressed mutant alleles or showed RNA interference (RNAi) for silencing endogenous genes have been designed and cell-wall components that were more easily converted to ethanol have been obtained (Chen & Dixon, 2007; Himmel et al., 2007). Examples of these strategies include the complementary decrease of lignin and the increase of cellulose components in cell walls or the directed overexpression of cellulases in plant cells to drastically decrease the cost of cell wall conversion to ethanol (Sticklen, 2008). However, the strategy involving lignin interference must be evaluated carefully in the context of biomass production because it could have side effects such as excessive sensitivity to fungal pathogens.

Because lignin is relatively resistant to enzymatic degradation, low-lignin transgenic trees have been investigated (Herrera, 2006). RNAi-mediated suppression of p-coumaroyl-CoA 3′-hydroxylase in hybrid poplars generally correlated very well with the reduction of lignin content. Up to ~13.5% more cell-wall carbohydrates have been observed in the suppressed lines as compared to wild-type poplars (Coleman et al., 2008).

Currently, lignocellulose pretreatment followed by enzymatic hydrolysis is the key process used for the bioconversion of lignocellulosic biomass (Sanderson, 2006). The type of pretreatment defines the optimal enzyme mixture to be used and the composition of the sugar mixture that is produced. Finally, the sugars are fermented with ethanol-producing microorganisms such as yeasts, Zymomonas mobilis, Escherichia coli, or Pichia stipitis (Fischer et al., 2008).

BIODIESEL

The Process of Biodiesel Production

The main components of plant oils are the fatty acids and their derivatives the mono-, diand triacylglycerides Tri-acylglycerides make up 95% of plant oils. Glycerides are esters formed by fatty acid condensation with tri-alcohol glycerol (propanetriol). Depending on the number of fatty acids fixed on the glycerol molecule, one can have mono-, di- or triacylglycerides. Of course, the fatty acids can be the same or different. As stated in the introduction, biodiesel can be obtained by esterification or transesterification. Esterification is the process by which a fatty acid reacts with a mono-alcohol to form an ester. The esterification reaction is catalyzed by acids. Esterification is commonly used as a step in the process of biodiesel fabrication to eliminate FFAs from low-quality oil with high acid content. Transesterification (or alcoholysis) is the displacement of alcohol from an ester by another alcohol in a process similar to hydrolysis. This process has been widely used to reduce triglyceride viscosity. The transesterification reaction is represented by the general equation (5).

RCOOR' + R"OH RCOOR" + R'OH (5)

This stepwise reaction occurs through the successive formation of di- and monoglycerides as intermediate products (Canakci et al., 2006). Theoretically, transesterification requires three alcohol molecules for one triglyceride molecule; however, an excess of alcohol is necessary because the three intermediate reactions are reversible (Marchetti et al., 2007; Om Tapanes et al., 2008). After the reaction period, the glycerol-rich phase is separated from the ester layer by decantation or centrifugation. The resulting ester phase (crude biodiesel) contains contaminants such as methanol, glycerides, soaps, catalysts, or glycerol that must be purified to comply with the European Standard EN 14214.

Different technologies can be used for biodiesel production; these include chemical or enzyme catalysis and supercritical alcohol treatment (Demirbas, 2008b). EN 14214 establishes 25 parameters that must be assessed to certify the biodiesel quality.

In conventional transesterification and esterification processes for the production of biodiesel, strong alkalis or acids are used as chemical catalysts. These processes are highly energy consumptive and the poor reaction selectivity that often results from the physicochemical synthesis justifies the ongoing research on enzymatic catalysis. In addition, an extra purification step is required to remove glycerol, water, and other contaminants from alkyl-esters.

The base catalysis is much faster than the acid catalysis. Low cost and favorable kinetics have turned NaOH into the most-used catalyst in the industry. However, soap and emulsion can be formed during the reaction and complicate the purification process.

Non-Edible Feedstocks for Biofuel Production

Currently, approximately 84% of the world biodiesel production is met by rapeseed oil. The remaining portions are from sunflower oil (13%), palm oil (1%), soybean oil and others (2%) (Gui et al., 2008). More than 95% of biodiesel is still made from edible oils. To overcome this undesirable situation, biodiesel is increasingly being produced from non-edible oils and waste cooking oil (WCO). Non-edible oils offer the advantage that they do not compete with edible oils on the food market.

Used cooking oil is a waste product, and for that reason, it is cheaper than virgin plant oil. The higher initial investment required by the acid-catalyzed process (stainless-steel reactors and methanol-distillation columns) is compensated for by low feedstock cost (Zhang et al., 2003). Reusing WCO esters provides an elegant form of recycling, given that waste oils are prohibited for use in animal feed, are harmful to the environment, and human health and disrupt normal operations at wastewater treatment plants (increasing the costs of both maintenance and water purification). The production of biodiesel from WCO is still marginal, but it is increasing worldwide. The USA and China are leaders in WCO use, with 10 and 4.5 Mt/yr, respectively. Other countries and regions, such as the EU, Canada, Malaysia, Taiwan and Japan, produce approximately 0.5-1 Mt/yr (Gui et al., 2008). The potential use of WCO as a primary source for biodiesel fuel is important because such use would negate most of the actual concerns regarding the competition of food and biodiesel crops for land (Bindraban et al., 2009; Odling-Smee 2007). By converting edible oils into biodiesel fuel, food resources are actually being converted into automotive fuels. It is believed that large-scale production of biodiesel fuels from edible oils may bring global imbalance to the food supply-and-demand market, even if such a trend has been contested (Ajanovic, 2010). However, nothing prevents the use of edible oils first for cooking and then for biodiesel fuel.

Biofuel Feedstocks in the World

Concerned by potential climate change-related damages (including changes to coastlines and the spread of tropical diseases, among others), the US faces the necessity of finding solutions for the 17.7%-reduction of GHG emissions (Lokey, 2007). Because of the fact that the electrical sector accounts for 40% of all GHG emissions, investments in cost-competitive renewable energy sources, such as wind, geothermal and hydroelectricity, have been recommended. Given the ample solar resources that exist in the US, it has a plethora of untapped sources for renewble-energy generation (Flavin et al., 2006). The Biomass Program of the US Department of Energy (launched in 2000) recommended 5% use of biofuels by 2010, 15% by 2017, and 30% by 2050. However, it is predicted that the ethanol market penetration for transportation should attain ~50% of gasoline consumption by 2030 (Szulczyk et al., 2010). Currently, maize and other cereals (such as sorghum) are the primary feedstocks for US ethanol production. At 40 Ml of ethanol per day, maize is still considered a low-efficiency biofuel crop because of its high required input, excessive topsoil erosion (10 times faster than sustainable) and other negative side effects (Donner & Kucharik, 2008; Laurance, 2007; Sanderson, 2006; Scharlemann & Laurance, 2008). By comparison, biodiesel from soybean requires lower inputs. However, neither of these biofuels can displace fossil fuel without impacting food supplies. Even if all US corn and soybean production were dedicated to biofuels, only 12% of the gasoline and 6% of the diesel demand, respectively, would be met (Hill et al., 2006). However, agricultural, municipal, and forest wastes could together sustainably provide 1 Gt of dry matter annually and should complement the other biofuel crops (Vogt et al., 2008). It was proposed that 3.1-21.3 Mha of land should be converted to biomass production (Schmer et al., 2008). Algal biodiesel is also being included in an integrated renewable-energy park (Singh & Gu, 2010; Subhadra, 2010).

Bioethanol from Brazil results in over 90% GHG savings (Hill et al., 2006). In addition to the PROALCOOL program, the Brazilian government created the PRO-ÓLEO program in 1980 and expected a 30% mixture of vegetable oils or derivatives in diesel and full substitution in the long term. Unfortunately, after the price drop of crude oil on the international market in 1986, this program was abandoned and was only reintroduced in 2002. Because of its great biodiversity and diversified climate and soil conditions, Brazil has a variety of plant-oil feedstocks, including mainly soybean, sunflower, coconut, castor bean, cottonseed, oil palm, physic nut and babassu (Nass et al., 2007). Brazil celebrated the inauguration of the Embrapa Agroenergia research center in 2010 to promote the integration of the oil from these feedstocks into the network of biodiesel sources. The National Program of Production and Use of Biodiesel (PNPB) was launched in 2004 with the objective of establishing the economic viability of biodiesel production together with social and regional development. The current diesel consumption in Brazil is approximately 40 Gl/yr and the potential market for biodiesel currently of 800 Ml and that should achieve 2 Gl by 2013. In addition, B5 has been mandatory since 2010. Auction prices have varied between US$ 0.3 and 0.8/l according to the area of production (Barros et al., 2006). Between 1975 and 1999, US$ 5 bn were invested in bioenergy resulting in the creation of 700,000 new jobs and US$ 43 bn saving in gasoline imports (Moreira & Goldemberg, 1999). The rate of job creation related to biodiesel production has been estimated to be 1.16 jobs/Ml of annual production (Johnston & Holloway, 2007). However, the recent trend of business centralization is expected to reduce this rate (Hall et al., 2009). Petrobras is now processing (with a capacity of 425,000 t) a mixture of plant oil and crude oil under the name of "H-Bio". With a tropical climate in the major part of its extention, the country has a potential 90 Mha that could be used for oleaginous crop production and that extends over Mato Grosso (southwest), Goiás, Tocantins, Minas Gerais (center), Bahia Piauí, and Maranhão (northeast).

The EU accounts for 454 million people (i.e., 7% of the world's population and 50% more people than live in the US) (Solomon & Banerjee, 2006). The EU is dedicated to a long-term www.intechopen. Conversion to a hydrogen economy. Renewable energy sources and eventually advanced nuclear power, are envisioned as the principal hydrogen sources on the horizon for use in 2020-250 (Adamson, 2004). However, even for the distant future, the EU foresees hydrogen production from fossil fuels with carbon sequestration still playing a major role (together with renewable energy and nuclear power). Because of their renewability, biodiesel and bioethanol in the EU have been calculated to result in 15–70% GHG savings when compared to fossil fuels. Frondel and Peters (2007) found that the energy and GHG balances of rapeseed biodiesel are clearly positive.

Bioethanol from sugar beets or wheat and biodiesel from rapeseed are currently the most important options available to the EU for reaching its target biofuel production. Because of increased land use for biofuel production, biofuel crops are now competing with food crops (Odling-Smee, 2007) and they are expected to have substantial effects on the economy. The European consumption of fossil diesel fuel is estimated to be approximately 210 Gl and that of biodiesel to be 9.6 Gl (Malça & Freire, 2011). The EU produces over ~2 Mha (i.e., ~1 Gl) of rapeseed (0.5 kl/ha) and sunflower (0.6 kl/ha) (Fischer et al., 2010), which shows that it depends heavily on importation of biofuels to approach the recommended target of B5.75. Given the higher energy potential of synfuel from biomass and the constraints on the availability of arable land, second-generation biofuels should soon enter the race for biofuel production (Fischer et al., 2010; Havlík et al., 2010).

The price for biodiesel that meets the EU quality standard (EN 14214) is approximately 730/t. By subtracting the biodiesel export value from the EU market price, one obtains the profit obtained by selling biodiesel from abroad on that market. The export value includes production and exportation costs. Production costs are made up of the plant oil or animal fat production plus the biodiesel processing minus the value of by-products (glycerol for example). Exportation costs include scaling, insurance, taxes and administrative costs (see the calculations in Johnston & Holloway, 2007). The price of US$ 0.88/l for biodiesel was 45% higher than the price of fossil diesel fuel during the same period (2006). Although this price is a convenient baseline, the biodiesel price on the EU market can change quickly depending on factors such as current domestic production, fossil diesel-fuel prices, agricultural yields, and legislation. The same rules will apply to emerging markets in China. Based on volume and profitability estimated in this manner, the top five countries that have the best combination of high volumes and low production costs are Malaysia, Indonesia, Argentina, the US, and Brazil. Collectively, these countries account for over 80% of the total biodiesel production. Plant oils currently used in biodiesel production account for only approximately 2% of global vegetableoil production, with the remainder going primarily to food supplies.

Despite the fact that India has not attained the high level of ethanol production seen in Brazil, it is the largest producer of sugar in the world. Indian ethanol is blended at 5% with gasoline in nine Indian states and an additional 500 Ml would be needed for full directive implementation. The total demand for ethanol is approximately 4.6 Gl (Subramanian et al., 2005). The country burns 3 times more fossil diesel fuel than gasoline (i.e., roughly 44 Mt), mainly for transportation purposes.

Because India imports 70% of its fuel (~111 Mt), any source of renewable energy is welcome. Therefore, India has established a market for 10% biodiesel blends (Kumar & Sharma, 2008). Because India is a net importer of edible oils, it emphasizes non-edible oils from plants such as physic nut, karanja, neem, mahua and simarouba. Physic nut and karanja are the two leaders on the Indian plant list for biodiesel production.

Of its 306 Mha of land, 173 Mha are already under cultivation. The remainder is classified as either eroded farmland or non-arable wasteland. Nearly 40% (80-100 Mha) of the land area is degraded because of improper land use and population pressures over a number of years. These wasted areas are considered candidates for restoration with physic nuts (Kumar & Sharma, 2008). Nearly 80,000 of India's 600,000 villages currently have no access to fuel or electricity, in part because there is not enough fuel to warrant a complete distribution network. Physic nuts could bring oil directly into the villages and allow them to develop their local economies (Fairless, 2007). This also applies to developing areas of Brazil and Africa.

In addition to the biodiesel initiative, regular motorcycles with 100 cm^3 internal combustion engines have been converted to run on hydrogen. The efficiency of these motorcycles has been proven to be greater than 50 km/charge. This development has had great significance because 70% of privately owned vehicles in India are motorcycles and scooters. Efforts are also underway to adapt light cars and buses to hydrogen, a move that will likely be helped by the growing number of electric and compressed natural gas (CNG) vehicles in and around New Delhi (Solomon & Banerjee, 2006).

In China, the area of arable land per capita is lower than the world's average. As a result, most edible oils are imported and the demand for edible oils in 2010 is projected to be 13.5 Mt. Because of its large population, China desperately needs sustainable energy sources. Because little arable land is available, China is exploring possibilities for the production of second- and third-generation biofuels (Meng et al., 2008). China is a large developing country that has vast degraded lands and that needs large quantities of renewable energy to meet its rapidly growing economy and accompanying demands for sustainable development. The energy output of biomass grown on degraded soil is nearly equal to that of ethanol from conventional corn grown on fertile soil. Biofuel from biomass is far more economic than conventional biofuels such as corn ethanol or soybean biodiesel. Potential energy production from biomass could reach 6,350,971 terajoules per year (TJ/yr) and an increased value of biomass in China's energy portfolio is considered unavoidable (Zhou et al., 2008).

Taking advantage of seawater availability, biodiesel from microalgae could also be conveniently grown along the 18,000 km Chinese coastline (Song et al., 2008). Marine microalgae production requires unused desert land, seawater, CO_2 and sunshine. Given the abundant areas of mudflats and saline lands in China, there is great potential to develop biodiesel production from marine microalgae.

Sales of electric bicycles and scooters in China have grown dramatically in the last 10 years and now total over 1 million per year. The growth of this demand has been facilitated by bans on gasoline-fueled bicycles and scooters in Beijing and Shanghai (among other large cities) because of increasing concerns about pollution (Solomon & Banerjee, 2006). For this reason, China has become one of the largest potential markets for hydrogen fuel cells in the transportation sector.

Frequent droughts in many Asian countries have made it difficult for them to replicate Brazil's success with sugarcane, which needs an abundant water supply. Thailand and Indonesia are tapping the potential with palm oil.

Because of its need to retain its position as the high-tech superpower for new technologies, Japan has become one of the most important players in the international development of a hydrogen-based economy. Following Japanese estimations, the hydrogen production potential from renewable energy in Japan is 210 GNm^3/yr(Nm^3 is the gas volume in m^3 at 0 °C and one atmosphere), which is 4 times more than what it will actually need in 2030. However, hydrogen based on renewable sources is only expected to contribute approximately 15% of the hydrogen consumed by 2030. It is estimated that on-board reforming of methanol or gasoline for fuel cell propelling would be the most practical technology in the near term, but the long-term goal is to adopt pure hydrogen (Solomon & Banerjee, 2006).

MICRODIESEL

Oleaginous microorganisms are microbes with an oil content that exceeds 20%. Biodiesel production from microbial lipids (known as single-cell oil or microdiesel) has attracted great attention worldwide. Although microorganisms that store oils are found among various microbes (such as microalgae, bacillus, fungi and yeast), not all microbes are suitable for biodiesel production (Demirbas, 2010).

Most bacteria are generally not good oil producers. Some exceptions are actinomycetes, which are capable of synthesizing remarkably high amounts of fatty acids (up to ~70% of their dry weight) from simple carbon sources such as glucose under growth-restricted conditions and which accumulate these fatty acids intracellularly as triglycerides (Alvarez & Steinbuchel, 2002).

The most efficient oleaginous yeast, Cryptococcus curvatus, can accumulate >60% lipids when grown under nitrogen-limiting conditions. These lipids are generally stored as triglycerides with approximately 44% percent saturated fatty acids, which is similar to many plant seed oils. Rhodotorula glutinis has been used for the wastewater treatment in monosodium-glutamate manufacturing. Monosodium-glutamate wastewater is as a cheap fermentation broth for the production of biodiesel using lipids from R. glutinis. To be efficient, the fermentation process needs a complementary source of glucose to obtain the proper C:N:P ratio (1:2.4:0.005). This process leads to a lipid production corresponding to 20% of the biomass after 72 h of culture and to an oil transesterification rate of 92% (Xue et al., 2008). In addition, R. glutinis can use various carbon sources including dextrose, xylose, glycerol, dextrose and xylose, xylose and glycerol, or dextrose and glycerol and can accumulate 16, 12, 25, 10, 21, and 34% triglycerides, respectively. The rate of unsaturated fatty acid accumulation was found to depend on the carbon source, with 25% and 53% accumulation when R. glutinis was grown on xylose and glycerol, respectively (Easterling et al., 2008). These results indicate that the use of R. glutinis can add value to several by-products, including glycerol. However, the resulting high levels of unsaturated fatty acids may require some additional saturation step to meet biodiesel standards.

Cyanobacteria are gram-negative photoautotrophic prokaryotes that can be cultivated under aqueous conditions ranging from freshwater to extreme salinity. They are able to produce a wide range of fats, oils, sugars and functional bioactive compounds such that their inclusion to wastewater treatment processes has been proposed (Markou & Georgakakis, 2011). Their duplication time is 3.5 h in the log phase of cell multiplication (Chisti, 2007). Using light energy, they are able to convert carbon substrates into oil (with a fatty acid composition that is similar to that of plants) at a rate of 20–40% of dry biomass (Meng et al., 2008).

Although microalgae are high-lipid-storing microbes, they require larger areas and longer fermentation times than do bacteria. The microalgae market produces approximately 5,000 t of dry biomass/year and generates approximately US$ 1.25 bn/yr (Pulz & Gross, 2004).

Eukaryotic diatoms, green algae and brown algae isolated from oceans and lakes typically reach dry-mass levels of 20%–50% lipids (Brennan & Owende, 2010). The quantities of lipids found in microalgae can be extraordinarily high. In Botryococcus, for instance, the concentration of hydrocarbons may exceed 80% of the dry matter. In comparison, drybiomass plant oil levels are generally around 15-40% lipids (Spolaore et al., 2006).

There are approximately 300 strains of algae, among which diatoms (including genera Amphora, Cymbella, and Nitzschia) and green algae (particularly genera Chlorella) that are the most suitable for biodiesel production. The oil is accumulated in almost all microalgaes as triglycerides (>80%) that are rich in C16 and C18 (Meng et al., 2008). Lipid accumulation in oleaginous microorganisms begins with nitrogen exhaust or when carbon is in excess (Ratledge 2002).

Chlorella protothecoides can accumulate lipids at a rate of 55% by heterotrophic growth under CO_2 filtration. Large quantities of microalgal oil have been efficiently recovered from these heterotrophic cells by n-hexane extraction. The microdiesel from heterotrophic microalgal oil obtained by acidic transesterification is comparable to fossil diesel and should be a competitive alternative to conventional biodiesel because of higher photosynthetic efficiency, larger quantities of biomass, and faster growth rates of microalgae as compared to those of plants (Song et al., 2008).

As stated above, microalgal oils differ from most plant oils in being quite rich in polyunsaturated fatty acids with four or more double bonds (Belarbi et al., 2000). This makes them susceptible to oxidation during storage and reduces their suitability for commercial biodiesel (Chisti, 2007). However, fatty acids with more than four double bonds can be easily reduced by partial catalytic hydrogenation (Dijkstra, 2006).

Changes in the degree of fatty acid unsaturation and the decrease or increase of fatty acid length are major challenges in modifying the lipid composition of microalgal oils. These features are regulated by enzymes that are mostly bounded to the cell membrane, which complicates their investigation (Certik & Shimizu, 1999). Currently, most of the genetic manipulations that have aimed to optimize metabolic pathways have been carried out on oleaginous microorganisms. This is mainly because of their abilities to accumulate high amounts of intracellular lipids, their relatively fast growth rates and their similarities of oil composition with plants (Kalscheuer et al., 2006a, 2006b).

Microalgae are often used for the sequestration and recycling of CO_2 by "CO_2 filtration" (Haag, 2007) and can reduce CO_2 exhaust by 82% on sunny days and by 50% on cloudy days (Vunjak-Novakovic et al., 2005). This process is much more elegant than carbon storage (CCS) in depleted oil fields or in aquifers because the carbon can be recycled via microdiesel. The storage capacity of CCS is estimated to range between 2,000-11,000 Gt CO_2; however, such aquifers are not evenly distributed around the world (Schiermeier et al., 2008). In addition, CCS does not result in any profit from the CO_2 that is stored and is actually an additional cost in the whole process. In contrast, algae convert CO_2 into oil. This means that the energy contained in the CO_2 can be re-injected into the power plant after being filtered by the algae and transformed into microdiesel.

The stimulation of fish production by increasing phytoplankton biomass through CO_2 injection into specific ocean localities has also been proposed (Markels & Barber, 2001). However, ocean fertilization has been severely challenged because it would eventually destroy the local ecosystem (Bertram, 2010; Glibert et al., 2008).

BIOHYDROGEN

The main alternative energy carriers considered for transportation are electricity and hydrogen. With interest in its practical applications dating back almost 200 years, hydrogen energy is hardly a novel idea. Iceland and Brazil are the only nations where renewableenergy feedstocks are envisioned as the major or sole future source of hydrogen (Solomon & Banerjee, 2006). Fuel-cell vehicles (FCVs) powered by hydrogen are seen by many analysts as an urgent need and as the only viable alternative for the future of transportation (Cropper et al., 2004).

Unlike crude oil or natural gas, reserves of molecular H_2 do not exist on earth. Therefore, H_2 must be considered more as an energy carrier (like electricity) than as an energy source (Song, 2006). H_2 can be derived from existing fuels such as natural gas, methanol or gasoline; however, the best long-term solution is to produce H_2 from water by (for example) using heat from solar sources and O_2 from the atmosphere.

Today, hydrogen is mainly manufactured by decarbonizing fossil fuels, but in the future it will be possible to produce hydrogen by alternative methods such as water photolysis using semiconductors (Khaselev & Turner, 1998) or by ocean thermal-energy conversion (Avery, 2002). Such methods are still in the research and development stage and are not yet ready for industrial application.

Hydrogen production from biomass requires multiple reaction steps. The reformation of fuels is followed by two steps in the water-gas shift reaction, a final carbon monoxide purification step and carbon dioxide removal.

Biomass can be thermally processed through gasification or pyrolysis. The main gaseous products resulting from the biomass are expressed by equations (6), (7) and (8) (Kikuchi, 2006).

$$\text{pyrolysis of biomass} \rightarrow H_2 + CO_2 + CO + \text{hydrocarbon gases} \quad (6)$$

$$\text{catalytic steam reforming of biomass} \rightarrow H_2 + CO_2 + CO \quad (7)$$

$$\text{gasification of biomass} \rightarrow H_2 + CO_2 + CO + N_2 \quad (8)$$

Hydrogen from organic wastes has generally been produced through equations (9), (10) and (11).

$$\text{solid waste} \rightarrow CO + H_2 \quad (9)$$

$$\text{biomass} + H_2O + \text{Air} \rightarrow H_2 + CO_2 \quad (10)$$

$$\text{cellulose} + 4H_2O + \text{Air} \rightarrow H_2 + CO + CH_2 \quad (11)$$

In the long run, the methods used for hydrogen production are expected to be specific to the locality. They are expected to include steam reforming of methane and electrolysis when hydropower is available (such as in Brazil, Canada and Scandinavia) (Gummer & Head, 2003). When hydrogen will become a very common energy source, it will likely be distributed through pipelines. Existing systems, such as the regional H_2-distribution network that has been operated for more than 50 years in Germany and the intercontinental liquid-hydrogen transport chain, demonstrate that leak rates of <0.1% can be achieved in industrial applications (Schultz et al., 2003). However, a major threat associated with the hydrogen paradigm is the fact that it is the smallest atom and that leakage is apparently unavoidable. One has to face the possibility that a significant amount of H_2 will be released into the stratosphere. Hydrogen is expected to react with ozone following the reaction $H_2+O_3 \rightarrow H_2O+O_2$. This mechanism (reviewed by Kikuchi, 2006) is a potentially dangerous promoter of ozone depletion. Alternatively, hydrogen can be produced from another fuel (e.g., ethanol, biodiesel, gasoline, or synfuel) via onboard reformers (hydrogen fuel processors). This is probably the best solution because synfuel can be produced from local feedstocks through the Fischer-Tropsch process, transported and distributed through existing technologies and infrastructures (Agrawal et al., 2007; Takeshita & Yamaji, 2008). This consideration also applies to biofuels. In addition, the feasibility of cars with onboard reformers has already been proven. The importance of synfuel is expected to increase rapidly because growing reserves of natural gas (or "stranded" gas) are available in remote locations and are considered to be too small for liquefied natural gas (LNG) or pipeline projects.

The biological generation of hydrogen (or biohydrogen) provides a wide range of approaches for generating hydrogen, including direct biophotolysis, indirect biophotolysis, photo-fermentation and dark-fermentation (Lin et al., 2010). Biological hydrogen production processes are found to be more environmentally friendly and less energy intensive as compared to thermochemical and electrochemical processes. There are three types of microorganisms that produce hydrogen, namely cyanobacteria, anaerobic bacteria, and fermentative bacteria (Demirbas, 2008a).

Photosynthetic production of H_2 from water is a biological process that can convert sunlight into useful, stored chemical energy. Hydrogen production is a property of many phototrophic organisms and the list of H_2 producers includes several hundred species from different genera of both prokaryotes and eukaryotes. The enzyme-mediating H_2 production seen in green algae is effected by a reversible hydrogenase that can catalyze ferredoxin oxidation in the absence of ATP (Beer et al., 2009). The enzyme is sensitive to oxidation; however, tolerant allozymes are being selected (Seibert et al., 2001). Hydrogen production has also been obtained from glucose using NADP+-dependent enzymes, glucose-6 phosphate dehydrogenase (G6PDH), 6-phosphogluconate dehydrogenase (6PGDH) and hydrogenase (Heyer & Woodward, 2001).

Carbon monoxide (CO) can be metabolized by a number of naturally occurring microorganisms along with water to produce H_2 and CO_2 following equation (12), which is the "water-gas shift" reaction, at ambient temperatures.

$$CO + H_2O \leftrightarrow CO_2 + H_2 \quad (12)$$

The biological water-gas shift reaction has been used in the processing of syngas from biomass with the bacterium Rubrivivax gelatinosus (Wolfrum & Watt, 2001). Nitrogenases can produce hydrogen but require relatively high energy consumption. However, the nitrogenase reaction is essentially irreversible, which allows for hydrogen pressurization. Rhodopseudomonas palustris can drive the nitrogenase reaction using light (Wall, 2004).

THE FUTURE OF TRANSPORT TECHNOLOGY

Fuel Cells

The fuel cell is the central component of hydrogen cars; it performs the conversion of fuel energy into electricity through proton mobilization. Fuel cells do not have moving parts, they produce only clean water and low-voltage electricity using hydrogen and oxygen, they are not noisy and they are 60% efficient, which is more than internal combustion engines (ICE, 45% efficiency). Laboratory tests indicate that fuel cells have a potential efficiency of 85% or more, which when combined with an 80%-efficient electric motor could make them 2 times more efficient than the direct use of hydrogen in an ICE (Ross, 2006).

Because of the security and cost problems related to infrastructure for hydrogen distribution and storage, ethanol is currently the most convenient alternative for fuel cells. Ethanol can be converted in hydrogen by onboard steam reforming or can be more conveniently used as a proton donor in specific fuel-cell technologies (Lamy et al., 2004). Ethanol-based steam reforming is performed following equation (13) (Velu et al., 2005).

$$C_2H_5OH + 3H_2O \rightarrow 2CO_2 + 6H_2 \qquad (13)$$

Deluga et al. (2004) described an onboard system for hydrogen production by auto-thermal reforming from ethanol. Following this system, ethanol and ethanol-water mixtures were converted directly into H_2 by catalytic oxidation with ~100% selectivity and >95% conversion and with a residence time on rhodium catalysts of <10 milliseconds. This process has great potential for low-cost H_2 generation in fuel cells for small portable applications in which liquid-fuel storage is essential and in which systems must be small, simple, and robust.

Another strategy of energy extraction from simple organic molecules is the glycerol biofuel cell (Arechederra et al., 2007). A biofuel cell is similar to a traditional proton exchange membrane (PEM) fuel cell. Rather than using precious metals as catalysts, biofuel cells rely on biological molecules (such as enzymes) to carry out the reactions. Arechederra et al. (2007) were able to immobilize two oxidoreductase enzymes (pyrroloquinoline quininedependent alcohol dehydrogenase and pyrroloquinoline quinine-dependent aldehyde dehydrogenase) at the surface of a carbon anode and to undertake a multi-step oxidation of glycerol into mesoxalic acid with 86% use of the glycerol energy. The bioanodes resulted in power densities of up to 1.21 mW/cm^2 using glycerol at concentrations up to 99 %. Because Nafion (the membrane) does not swell under glycerol, the biofuel cell longevity is expected to be higher than the technology used at moment.

Formula 1 has entered the race for optimizing green technologies. From 2009 on, new regulations for Formula 1 have forced the racing teams to recover the energy lost in braking and to use it to propel the car (Trabesinger, 2007). The technology that accomplishes this is called a "kinetic-energy recovery system" (KERS, better known as "regenerative braking"). In a hybrid car with both combustion and electric motors, batteries can be charged either by the ICE or by regenerative braking. The stored electric energy is then used to power the car at low speeds (i.e., in the city traffic) where the ICE efficiency is low because of continuous "stop-and-go" motion.

Fuel cells are still very expensive and currently cost approximately US$ 4,000/kW, which is 100 times more expensive than the cost of ICEs. Fuel-cell stacks must be replaced 4–5 times during the lifetime of current generations of vehicle. It is thus the cost of 4–5 fuel-cell units that must be compared with alternative ICEs (Marcinkoski et al., 2008; Sorensen, 2007).

Therefore, to be competitive with ICEs, the technology must reach the threshold of US$ 30/kW. To address this situation, Honda is selling its first prototype fuel-cell car under a leasing contract in California. BMW has been a pioneer of fuel-cell technology and produced its first hydrogen-car prototype in the 1960s (Hissel et al., 2004). Its current vehicle uses liquid hydrogen with autonomy of up to 386 km. The Ford Motor Company has set a new land-speed record for a fuel-cell powered car (334 km/h).

Despite these pilot experiments, it is likely that urban buses will be among the first large scale commercial applications for fuel cells. This is due to the fact that urban buses are highly visible to the public, contribute significantly to air and noise pollution in urban areas, have few size limitations and are fueled via a centralized infrastructure. Folkesson et al. (2003) reported the following: (i) the net efficiency of a Scania bus powered by a hybrid PEM fuel-cell system was approximately 40%; (ii) the fuel consumption of the hybrid bus was between 42 and 48% lower than that of a standard ICE Scania bus; and (iii) regenerative braking saved up to 28% energy. The bus prototype was equipped with a fuel cell of 50 kW and was fueled with compressed ambient air and compressed hydrogen stored on the roof. All of the fossil fuel options result in large amounts of GHG emissions. Ethanol and hydrogen have the potential to significantly reduce greenhouse gas emissions. However, their use will be highly dependent on pathways of ethanol and hydrogen production. Some of the hydrogen options result in higher GHG emissions than do ICEs running on gasoline. The vehicle options that will be competitive during the next two decades are those that use improved ICEs (including hybrids burning 'clean' gasoline or diesel). In the present state of the technology, cars running on hydrogen using onboard reforming of carbon fuel are still ecologically less efficient than are gasoline ICEs. The relatively high energy consumption required to produce hydrogen is expected to affect the geographic distribution of hydrogenpowered cars. One can speculate that such cars would be more appropriate in areas where solar (Eugenia Corria et al., 2006), wind or hydro-electricity power sources are abundant.

Energy Storage

A variation on the hybrid vehicle is the 'plug-in hybrid', which can be connected to the electric grid. The savings in energy costs over the whole cycle of charging an onboard battery and then discharging it to run an electric motor in an electric-hybrid (e-hybrid) car is 80%. This figure is approximately 4 times higher than the savings from fuel-cell cars running on hydrogen made using electrolysis and 30% higher than savings from cars running on gasoline (Romm, 2006). These vehicles allow the replacement of a substantial portion of the fuel consumption and tailpipe emissions. If the electricity is produced from CO_2-free sources, then e-hybrids can also have dramatically reduced net greenhouse gas emissions.

The electrical storage system is the key element of the e-hybrid car because its power capacity and lifetime decisively define the costs of the overall system (Bitsche & Gutmann, 2004).

Bio-based energy-management processes are emerging and could make a significant contribution in the medium term. The production of electricity is also possible with whole microorganism fermentation Fe(III)-reducing microorganisms in the family Geobacteraceae can directly transfer electrons onto electrodes (Bond et al., 2002; Bond & Lovley, 2003). However, the range of electron donors that these organisms can use is limited to simple organic acids. By contrast, Rhodoferax ferrireducens is capable of oxidizing glucose and other sugars (such as fructose and xylose) with similar efficiency and of quantitatively transferring electrons to graphite electrodes. The sugar is consumed in the anode chamber. The oxidation of one molecule of glucose produces CO_2, H+ and 24 electrons with a ~83% efficiency. The reaction produces a long-term steady current that is sustained after glucosemedium refreshing in the anode chamber. This microbial fuel cell can be recharged by changing the anode medium. It does not show severe capacity fading in the charge/discharge cycling and only presents low-capacity losses under open circuits and prolonged idle conditions (Chaudhuri & Lovley, 2003

Another bacterium that is able to transfer electrons to solid metal oxides is Shewanella oneidensis MR-1. In addition, to their remarkable anaerobic versatility, analyses of the genome sequences of Shewanellae species suggest that they can use a broad range of carbon substrates; this creates possibilities for their application in biofuel production (Fredrickson, 2008). Production and storage of electricity are expected to evolve quickly within the new paradigm of emerging bioelectronics (Willner, 2002) sol-gels have been demonstrated to be usable for the entrapment of membrane-bound proteins in a physiologically active form and have been proven to be capable of maintaining protein activity over periods of months or more (Luo et al., 2005). Using a membraneassociated F0F1-ATP synthase, Luo et al. (2005) showed that the photo-induced proton gradient can be used to 'store' light energy as ATP. This has the advantage of eliminating passive leakage of ions across the membrane. In addition, ATP can be used for direct powering of motor proteins for the conversion of chemical energy to mechanical energy (Browne & Feringa, 2006). Nano power plants based on the rotation of magnetic bead propellers mounted on F_0F_1-ATP synthase rotors that are fed by ATP to induce electric current in microarrays of nanostators are now being designed and are in the research and development stage of construction (Soong et al., 2000; Yasuda et al., 2001).

Options for Grid Contributions

Electricity is the foundation of modern societies, yet more than 1.6 billion people remain without access to the electrical grid. A majority of this population lives in South Asia and sub-Saharan Africa. Despite global economic expansion and advances in energ technologies, roughly 1.4 billion people (or 18% of the world's population) will still b without power by 2030 unless major governmental incentives are put into place (Dorian et a!., 2006)

The world average annual electricity consumption is between 2 and 4 TW. The cost of fossil derived electricity is now in the range of US$ 0.02–0.05/kW/hr, including storage and distribution costs (Lewis & Nocera, 2006). For comparison, the options of non-biological electricity generation are as follows. (i) The light-water reactors that make up most of the world's nuclear capacity produce electricity at costs of US$ 0.025-0.07/kW/; however, there is no consensus as to the solution to the problem of how to deal with the nuclear wastes that have been generated in nuclear power plants over the past 50 years (Schiermeier et al., 2008). (ii) Hydroelectric energy sources have a generating capacity of 800 GW (i.e., 10 times more power than geothermal, solar and wind power sources combined) and currently supply approximately one-fifth of the electricity consumed worldwide. Annual operating costs are US$ 0.03-0.10/kW/h, which makes such sources competitive with coal and gas. Because only approximately 30% of worldwide hydroelectric capacity is currently used, energy from these sources can still be tripled (Schiermeier et al., 2008). (iii) Wind turbines can produce 1,500 kW at US$ 0.05-0.09/kW/h making wind competitive with coal; wind power could provide up to 20% of the electricity in the grid. The EU should be able to meet25% of its current electricity needs by developing wind power in less than 5% of the North Sea and is heavily investing in that option. (iv) Exploitation and resulting use of the best geothermal sites is estimated to cost approximately US$ 0.05/kW/h. Thus, 70 GW of the global heat flux is seen as exploitable. However, because of the great deal of investment required exploitation of geothermal power lies outside of current priorities except in regions with significant volcanic activity (Schiermeier eta!., 2008). (iv) Commercial photo voltaic (PV) electricity costs US$ 0.25-0.30/kW/h, which is still 10 times more than the current price of electricity on the grid.

The possibility for use of current PV technology is limited to 31% by theoretical considerations. A conversion efficiency of >31% is possible if photons with high energies are converted to electricity rather than to heat. With use of such technology, the conversion efficiency could be >60% (Lewis, 2007). The absence of a cost-effective storage method for solar electricity is also a major problem. Currently, the cheapest method of solar-energy capture, conversion, and storage is solar thermal technology, which can cost as little as US$ 0.10-0.15/kW/h for electricity production. This requires the focusing of the energy in sunlight for syngas or synfuel synthesis (Lewis & Nocera, 2006) or its thermal capture by heat-transfer fluids that are able to sustain high temperatures (>427 °C) and resulting electricity generation through steam production (see in Shinnar & Citro, 2006). Solar power is among the most promising carbon-free technologies available today (Schiermeier et al., 2008). The earth receives approximately 100,000 TW of solar energy each year. There are areas in the Sahara Desert, the Gobi Desert in central Asia, the Atacama in Peru and the Great Basin in the US that are suitable for the conversion of solar energy to electricity. The total world energy needs could be fed using solar energy captured in less than a tenth of the area of the Sahara. Residential and commercial roof surfaces are already being used in several countries to allow the people to sell their own PV electricity to the grid (and in this way saving substantial annual costs). This elegant strategy could be extended to other systems of energy production.

The capital costs of biomass are similar to those of fossil fuel plants. Power costs can be as little as US$ 0.02/kW/h when biomass is burned with coal in a conventional power plant. Costs increase to US$ 0.04-0.09/kW/h for a co-generation plant, but the recovery and use of the waste heat makes the process much more efficient. The biggest problem for new biomass power plants is finding a reliable and concentrated feedstock that is available locally. Biomass production is limited by land-surface availability, the efficiency of photosynthesis, and the water supply. Biomass potential is estimated at ~5 TW (Schiermeier et al., 2008). Photosynthesis is relatively inefficient if one considers that in switchgrass (one of the fastestgrowing crops), energy is stored in biomass at an average rate of <1 $W/m^2/yr$. Given that the average insolation produces 200-300 W/m^2, the average annual energy conversion and storage efficiency of the fastest growing crops is only <0.5% (Lewis 2007; Lewis & Nocera, 2006). However, photosynthetic efficiency can be improved by genetic engineering (Ragauskas et al., 2006). Another potential problem with biomass production is that it could result in an increase of water consumption of two to three orders of magnitude. This is an important consideration because basic human necessities and power generation are increasingly competing for water resources (King et al., 2008).

The potential availability of wind (Pryor & Barthelmie, 2010), solar and biomass energy varies over time and location. This variation is not only caused by the individual characteristics of each resource (e.g., wind and solar regimes, soils), but also by geographic (land use and land cover), techno-economic (scale and labor costs) and institutional (policy regimes and legislation) factors (de Vries et al., 2007). The regional potential in energy units/year must be integrated over the geographical units that belong to a particular region. The model from de Vries et al. (2007) showed the following: (i) electricity from solar energy is typically available from Northern Africa, South Africa, the Middle East, India, and Australia; (ii) wind is concentrated in temperate zones such as Chile, Scandinavia, Canada, and the USA; (iii) biomass can be produced on vast tracts of abandoned agricultural land typically found in the USA, Europe, the Former Soviet Union (FSU), Brazil, China and on grasslands and savannas in other locations. In many areas of India, China, Central America ,South Africa and equatorial Africa, these energy sources are available at costs of below US$ 0.1/kW/h and are found in areas where there is already a large demand for electricity (or there will be such demand in the near future). A combination of electricity from wind, biomass and/or solar sources (Eugenia Corria et al., 2006) may yield economies-of-scale in transport and storage systems. Regions with high ratios of solar-wind-biomass potential to current demand for electricity include Canada (mainly wind), African regions (solar-PV and wind), the FSU (wind and biomass), the Middle East (solar-PV) and Oceania (all sources). In other region (such as Southeast Asia and Japan), the solar-wind-biomass supply is significantly lower than the demand for electricity. Ratios of around one are found in Europe and South Asia. The potentials just described depend on many parameters, and their achievement will depend on future land-use policies (de Vries et al., 2007; Miles & Kapos, 2008).

MANAGEMENT AND SUSTAINABILITY

Adam Smith's notion that by pursuing his own interest a man "frequently promotes that of society more effectively than when he really intends to promote it" and Karl Marx's picture of a society in which "the free development of each is the condition for the free development of all" are both limited by one obvious constraint. The world is finite. This means that when one group of people pursues its own interests, it damages the interests of others (Vertès et al., 2006). The model of Western economies was established using this logic. The theoretical framework of this philosophy is a mathematical model that is based on energy-conservation equations formulated by von Helmholtz in 1847, in which physical variables were arbitrarily substituted by economic ones. The consequences of this model are as follows: (i) the market is a closed circular flux between production and consumption, without inflows or outflows; (ii) natural resources are located in a domain that is separate from that of the closed market system; (iii) the costs of environmental destruction because of economic activities must be considered as unrelated to the closed market system (or at least they cannot be included in the price-formation processes of that system); (iv) the natural resources that are used by the market system are endless and those that are limited in quantity can be substituted by others that are endless; and (v) biophysical limits to the increase of the market system simply do not exist (Nadeau, 2006). This model is obsolete and is based on hypotheses that have no grounding in scientific bases. Sustainable economic solutions to global warming and environmental destruction are impossible to establish under the logic of this model.

As a consequence, the US alone has reached a level of oil consumption in the transportation sector that approaches 14 Mbl/day and corresponds to a release of 0.53 gigatons of carbon per year (Gt C/yr). The current global release of carbon from all fossil fuel usage is estimated to be at 7 Gt C/yr and is expected to rise to ~14 Gt C/yr by 2050 (Agrawal et al.,2007). It has been estimated that global energy consumption could reach 30-60 TW by 2050.With world population expected to reach 8 billion by 2030, the scale-up in energy use that is needed to maintain economic growth is critical. China, with 1.3 billion people and a fast growing economy, has overtaken Japan to become the second-largest oil consumer behind the US. The Asian giant is currently the largest producer and consumer of coal (Tollefson, 2008) and has announced the construction of 24-32 new nuclear reactors by 2020 (Dorian et al., 2006). If current trends continue, the world will need to spend an estimated $16 trillion over the next three decades to maintain and expand its energy supply. Generation, transmission, and distribution of electricity will absorb almost two-thirds of this investment, whereas capital expenditures in the oil and gas sectors will amount to almost 20% of global energy investment.

Experts believe that peak of world oil production should not occur before at least 30-40 years from now. To put global oil needs into perspective, demand for oil is projected to rise from nearly 80

Mbl/day today to over 120 Mbl/day by 2030. The OPEC nations are currently operating at near full capacity, which caused oil prices to reach US$ 120/bl in August 2008. Clearly, the world must find more efficient ways to manage energy. Some argue that the supplies of oil needed to satisfy the growing world demand will become available because of a combination of price and technology incentives (Rafaj & Kypreos, 2007). As oil prices continue to rise because of increasing difficulties in reaching remaining oil resources, other energy forms will appear (Herrera, 2006). A transition from oil to renewable energy should occur at some point before the world runs out of oil resources (Dorian et al., 2006). Renewable energy sources, including solar, wind, and geothermal, but excluding biofuels, currently provide only 3% of world energy demand (Dorian et al., 2006). Solutions that use these energy sources should be increased worldwide and should be connected to the electricity grid.

Renewable biodiesel from palm oil and bioethanol from sugarcane are currently the two leaders of plant bioenergy production per hectare. They are being grown in increasing amounts; however, continuous increases in their production are not sustainable and will not resolve the enormously increasing demands for energy. Palm oil yields ~5,000 l/ha. In Brazil, the best bioethanol yields from sugarcane are 7,500 l/ha. Most of the energy needed for growing the sugarcane and converting it to ethanol is gained from burning its wastes (e.g., bagasse). For every unit of fossil energy that is consumed by producing sugarcane ethanol, ~8 units of energy are recovered (Bourne, 2007). The rates of energy recovery from other biofuel crops are usually less than 5. Biofuel crops from the EU are much less productive than palm oil and sugarcane; therefore, B5 enforcement would require that ~13% of the EU25 arable land be dedicated to biofuel production. This is hardly sustainable (the present situation is ~5 times less).

Regarding environmental impact, ethanol from corn (for example) contains costs that stem from the copious amounts of nitrogen fertilizer used and the extensive topsoil erosion associated with cultivation. Every year, pesticides, herbicides and fertilizers run off the corn fields and bleed into groundwater. River contamination promotes eutrophication, algal blooms and' dead zones'. In addition, ethanol importation by industrialized nations could lead to increased ecological destruction in developing countries as indigenous natural habitats are cleared for energy crops (Gui eta!., 2008; Marris, 2006; Thomas 2007).

The general feeling is that first-generation biofuels are already reaching saturation because of the limited availability of arable lands. Brazil has additional lands available for sugarcane and physic nut production, whereas India is promoting physic nut cultivation on its extensive wastelands. However, the development of these fuels has already been a success because they have demonstrated that motor technology running on ethanol or biodiesel is feasible and can (at least) be used to power public transport.

Fortunately, second-generation biofuels from biomass offer additional opportunities. The cost of feedstock is lower for lignocellulose as compared to the agricultural crops that now contribute up to 70% of the total production costs for first-generation bioethanol. Even if they are more expensive now, synfuel from biomass sources (such as poplar, willow, and reed grass) could have higher cost effectiveness in the near future than does fuel from sugar beets, wheat and rapeseed sources (Wesseler, 2007; Styles & Jones, 2008).

Biomass fuels will be another opportunity for the EU to meet its target of energy production from renewable sources. However, this goal has not been met by 2010 as was initially expected (Fischer et al., 2010; Havlík et al., 2010). The European CO_2 emissions-trading system of carbon credits seems to be much more cost effective than its biodiesel program because it allows for the purchase of units of CO_2 sequestration in tropical climates that have much higher rates of fixation than do temperate ones (Frondel & Peters, 2007).

Third-generation biofuels have also entered the race for fuel renewability. In terms of total dry matter, sugarcane typically yields ~75 t biomass per hectare, whereas micro*algae* are able to produce two times more biomass per hectare (Brennan & Owende, 2010; Chisti, 2007, 2008). Considering a productivity of 150 t/ha and an average dry-weight oil content of 30%, the oil yield per hectare would be ~123 m3 over 90% of the year (i.e., 98.4 m3/ha). If 0.53 Gm3 of biodiesel are needed in the US to power transport vehicles, micro*algae* should be grown over an area of ~5.4 Mha (3% of the US). Producing algal biomass in a 100 t/yr facility has been estimated to cost approximately US$ 3,000/ton. The feasibility of oil extraction for microalgal biomass has been demonstrated (Belarbi et al., 2000; Sánchez Mirón et al., 2003) and the majority of algal biomass residues from oil extraction can be recycled by anaerobic digestion to produce biogas.

Impediments to large-scale culture of micro*algae* are mainly economic and are tied to the investment requirements for the *algae* cultivation. One solution would be to increase the oil productivity by genetic and metabolic engineering (León-Bañares et al., 2004; Mathews & Wang, 2009). One may expect the expansion of algal technology via CO_2 filtration because power plants can incorporate this technology immediately into their management systems. This technology is expected to spread slowly with the accumulation of experience.

Nearly half of the world's oil consumption is dedicated to the transportation sector, which also accounts for 32% of GHG emissions. The overall efficiency of energy conversion to goal is to increase it from the current level of 15–35 to 60–80% (Song, 2006).

Unfortunately, advanced transportation technologies (such as hydrogen fuel cell vehicles and alternative fuels including gas-to-liquids, coal-to-liquids, and biodiesels) are not likely to significantly penetrate the conventional transportation fuel market before 2030 (except on a regional basis). The growth in oil consumption for transportation use in the coming decades may be slowed by the adoption of fourth-generation technologies such as hybrids and fuel cell cars. However, the necessary technological breakthroughs will not occur without unprecedented policy actions worldwide to promote the use and inclusion of these technologies in everyday life (Doniger et al., 2006; Haug et al., 2011; Michel 2009). Currently, there are approximately half a million hybrids and 30 million advanced clean-diesel engines globally. The use of hybrid cars is growing in the US and Japan, whereas advanced cleandiesel motors are mostly concentrated in Europe (Dorian et al., 2006). Actually, auto-mobility is a self-organizing and non-linear system that presupposes and calls into existence an assemblage of cars, drivers, roads, fuel supplies, and other objects and technologies. Modern social life has become interconnected with auto-mobility. However, this mode of mobility is neither socially necessary nor inevitable (Urry, 2008). One billion cars were produced during the last century. World automobile travel is predicted to triple between 1990 and 2050 (Hawken et al., 2002). Today, world citizens move 23 Gkm annually. Auto-mobility forces people to contend with the temporal and spatial constraints that it itself generates (Mills et al., 2010). Fortunately, some 35-year-old projects have begun to be finally implemented (i.e., the integration of car and bicycle rentals into public transportation systems, such as occurs in some European cities). A post-car future will involve changes in lifestyles, city architecture, thinking and social practices. Increased active transport (e.g., walking and bicycling) will help to achieve substantial reductions in emissions while improving public health. Cities require safe and pleasant environments for active transport as well as easy accessibility of public transport. Adverse health effects because of transportation include traffic injuries, physical inactivity (the cost of obesity in the USA is estimated to be around US$ 139 bn/yr), urban air pollution, energy-related conflicts, and environmental degradation. For instance, urban air pollution accounts for 750,000 deaths each year, of which 530,000 are in Asia (Woodcock et al., 2007). Because of limited energy resources, it has been argued that the world will be required to move toward virtual travel (such as internet surfing, virtual sensorial traveling, and video conferences) to replacephysical travel as much as possible (Moriarty & Honnery, 2007).

In reality, the situation outlined above is the result of consideration of humanity only within social contexts and without the necessary environmental perspective (Thomas, 2007). The concept of environmental crime barely operational; if it exists at all, it is very recent and is not generally applied. Logical human societies should take into account the amount of land that human beings and wildlife actually need to reasonably sustain themselves. Not doing this will lead to increasing worldwide destruction (Urry, 2008) and will threaten the future of humanity. These considerations led to the formulation of the Gaia principle (Lovelock & Margulis, 1974). This principle states that one should consider the planet Earth as a whole, with the consequence that the destruction of one ecosystem can affect all of the others. Concern for the value of ecosystems is recent (Costanza et al., 1997). Society has only begun to address human integration with the environment because of the threat of global warming and its potentially disastrous effects (Stern, 2006). A discussion of the economic accounting for ecosystem services from the perspective of sustainable development has also been proposed (Mäler et al., 2008).

The concept of "willingness-to-pay" (WTP) has also been recently introduced. This concept allows for the monetary measurement of individual preference to avoid a negative impact. It aims to estimate the need for improved environmental quality. WTP measures how much individuals are ready to pay to improve their quality of life or that of other people. The sum of the WTP of all individuals gives the value that a group of individuals are ready to pay to maintain their environment in an unaffected state. For example, the pathways of polluting substances are followed from their release sources to the points of damage occurrence with associated "external" costs of reparation. Taking external costs into account in the full cost of energy production leads to the estimation of the "real" cost of an activity and supplies an efficient policy instrument for reducing the negative impacts of energy use (Nast et al., 2007). The approach of merging production costs with external costs into a total specific cost serves as a comparative indicator for the evaluation of the economic-environmental performance of energy options and technologies (Rafaj & Kypreos, 2007). The scenarios proposed under this new cost-accounting strategy reveal the possibilities for the diffusion of advanced technologies and fuel switching into the electricity production system. Following this model, renewable energies increase their competitiveness and the dependency of the electricity sector on fossil fuels is decreased considerably. Additionally, emissions of SO2 and NOx decrease by 70–85% by 2030. Although the analysis indicates that advanced technologies with emission controls and carbon sequestration will undergo significant cost reduction and will become competitive in the long run, policies supporting these technologies are aprerequisite to their establishment in electricity markets (especially during their initial period of market penetration). This model refers to policy measures for the stimulation of technological progress via investments in research and development that assist carbon-free technologies to progress along their necessary learning curves (Haug et al., 2011; Rafaj & Kypreos, 2007).

CONCLUSIONS

The time has come for the integration of the technological and social sciences to find a route to environmental and economic sustainability on earth. If such a solution is not reached, economic growth will occur at the cost of the human population size (Urry, 2008). Fortunately, because of the continuous increase in the price of fossil fuel, investigations intosources of renewable energy have become economically viable. It is now clear that technologies for renewable energies have reached a pivotal stage such that there is no turning back. There are at least 5 regional blocks (the USA, the EU, China, Brazil, and India) that are interested in decreasing their dependence on fossil fuels. It does not appear to be in anyone's interest to shut this process down by mean of aggressive oil price cutting and market dumping. In fact, biotechnology is intimately bound to agricultural processes that are also supported by governments because of geostrategic issues. In addition, climate change is becoming obvious and will soon overcome particular interests to become a general concern of humanity.

Biofuels and sources of bioenergy will pass through a rapid succession of technological improvements and developments before they arrive in their final forms. It is expected that bioethanol from sweet crops will be surpassed by bioethanol from biomass. Synfuel from biomass and solar energy should also progressively replace plant biodiesel. Biotechnology is expected to increase its participation in microdiesel fuel production, in genetic engineering of plants and microorganisms and in the contribution of enzymes to nanotechnology.

The integration of renewable energies into the electricity grid is just beginning, but is already progressing rapidly. It is expected to make a significant contribution; however, it should be accompanied by policies of energy management and urbanization to avoid unnecessary energy waste that could negate the benefits of technological breakthroughs and developments. New concepts (such as willingness-to-pay, carbon credits and external costs) are now being taken into account in the calculation of energy life cycles. This toolbox will expand with increasing government regulations and should include fundamental concepts such as "biodiversity credits" and the definition of a "minimal territorial unit" for living entities to warrant sustainability of wildlife and humanity. Biodiversity is a source of nanostructures and nanomachines. It should not be destroyed without consideration when we are aware that it required three billions years to develop and that humanity is just beginning to investigate it.

As a result of energy saving requirements, the cars of the near future will run on combinations of fuel combustion and electricity. Such options can reduce fossil fuel consumption and greenhouse gas emissions by 30 to 50%, with no gross vehicle modifications required. In addition, they will allow for connection to the electricity grid for additional cost saving on electricity consumption. These so-called plug-in hybrids will likely travel three to four times farther per kW/h than other vehicles. Ideally, these advanced hybrids will also be flexible and capable of running on bio/fossil blends and gas (Romm, 2006). At some point during the first half of this century, a transition from fossil fuels to a noncarbon- based world economy will begin and will seriously affect the type of society experienced by future generations (Dorian et al., 2006).

ACKNOWLEDGEMENTS

We thank the Coordenação de Aperfeiçoamento de Pessoal de Nível Superior (CAPES) and Fundação Oswaldo Cruz (FIOCRUZ) for providing a research fellowship from the Centro de Desenvolvimento Tecnológico em Saúde (CDTS) to N. Carels. This work received financial support from Conselho Nacional de Desenvolvimento Científico e Tecnológico (CNPq), Brazil (no. 471214/2006-0).

REFERENCES

1. Abollé, A., Kouakou, L. & Planche, H. 2008. The viscosity of diesel oil and mixtures with straight vegetable oils: Palm, cabbage palm, cotton, groundnut, copra and sunflower. *Biomass Bioenergy* 33:1116-1121.
2. Adamson, K.A. 2004. Hydrogen from renewable resources - the hundred year commitment. *Energ. Policy* 32:1142–1231.
3. Agarwal, A.K. 2007. Biofuels (alcohols and biodiesel) applications as fuels for internal combustion engines. *Prog. Energ. Combust. Sci.* 33: 233-271.
4. Agrawal, R., Singh, N.R., Ribeiro, F.H. & Delgass, W.N. 2007. Sustainable fuel for the transportation sector. *Proc. Natl. Acad. Sci. USA* 104:4828-4833.
5. Agrawal, A., Chhatre, A. & Hardin, R. 2008. Changing Governance of the World's Forests. *Science* 320:1460-1462.
6. Ajanovic, A. 2011. Biofuels versus food production: Does biofuels production increase food prices? *Energy*, doi:10.1016/j.energy.2010.05.019
7. Alvarez, H.M. & Steinbuchel, A. 2002. Triacylglycerols in prokaryotic microorganisms. *Appl. Microbiol. Biot.* 60:367-376.

8. Arechederra, R.L., Treu, B.L. & Minteer, S.D. 2007. Development of glycerol/O2 biofuel cell. *J. Power Sources* 173:156-161.
9. Arthur, R., Baidoo, M.F. & Antwi, E. 2011. Biogas as a potential renewable energy source: A Ghanaian case study. *Renew. Energ.* 36:1510-1516.
10. Asam, Z.-u.-Z. 2011. How can we improve biomethane production per unit of feedstock in biogas plants? *Appl. Energ.*, doi:10.1016/j.apenergy.2010.12.036
11. Avery, W. 2002. Ocean-thermal energy conversion. In: Encyclopedia of Physical Science and Technology, edited by Meyers, R.A., 123- 160. San Diego: 3rd ed. Academic Press. Baker, E. & Keisler, J.M. 2011. Cellulosic biofuels: Expert views on prospects for advancement. *Energy* 36:595-605.
12. Barnet, A. 2008. Europe's 2020 vision. *Nat. Rep. Clim. Change* 2:36.
13. Barros, G.S.A.C., Silva, A.P., Ponchio, L.A., Alves, L.A., Osaki, M. & Cenamo, M. 2006. Custos de produção de biodiesel no Brasil. *Revista de Política Agrícola* 3:36-50.
14. Beer, L.L., Boyd, E.S., Peters, J.W. & Posewitz, M.C. 2009. Engineering *algae* for biohydrogen and biofuel production. *Curr. Opin. Biotech.* 20:264–271.
15. Begum, Z.N. 2005. Modeling of chemical tracer transport in the atmospheric environment and its impact on the global climate. *J. Quant. Spectrosc. Radiat. Transfer* 95:423-427. Belarbi, E.-H., Molina Grima, E. & Chisti, Y. 2000. A process for high yield and scaleable recovery of high purity eicosapentaenoic acid esters from micro*algae* and fish oil. *Enzyme Microb. Tech.* 26:516-529.
16. Bertram, C. 2010. Ocean iron fertilization in the context of the Kyoto protocol and the post- Kyoto process. *Energ. Policy* 38:1130–1139
17. Bindraban, P.S., Bulte, E.H. & Conijn, S.G. 2009. Can large-scale biofuels production be sustainable by 2020? *Agr. Syst.* 101:197–199
18. Bitsche, O. & Gutmann, G. 2004. Systems for hybrid cars. *J. Power Sources* 127:8–15.

19. Bond, D.R., Holmes, D.E., Tender, L.M. & Lovley, D.R. 2002. Electrode-reducing microorganisms that harvest energy from marine sediments. *Science* 295:483-485.
20. Bond, D.R. & Loveley, D.R. 2003. Electricity production by *Geobacter sulfurreducens* attached to electrodes. *Appl. Environ. Microb.* 69:1548-1555.
21. Bourne, J.K.Jr. 2007 Biofuels: green dreams. *National Geographic Magazine* (October), 41–59.
22. Brennan, L. & Owende, P. 2010. Biofuels from micro*algae* - A review of technologies for production, processing, and extractions of biofuels and co-products. *Renew. Sustain. Energ. Rev.* 14:557-577
23. Browne, W.R. & Feringa, B.L. 2006. Making molecular machines work. *Nature Nanotechnology* 1:25-35.
24. Canakci, M., Erdil, A. & Arcaklioglu, E. 2006. Performance and exhaust emissions of a biodiesel engine. *Appl. Energ.* 83, 594-605.
25. Certik, M. & Shimizu, S. 1999. Biosynthesis in microorganisms being used for single cell oil production. *Biochimie* 87:1-14.
26. Ceschia, E. et al. 2010. Management effects on net ecosystem carbon and GHG budgets at European crop sites. *Agricult. Ecosys. Environ.* 139:363–383.
27. Chapple, C., Ladisch, M. & Meilan, R. 2007. Loosening lignin's grip on biofuel production. *Nat. Biotechnol.* 25:746-748.
28. Chaudhuri, S.K. & Lovley, D.R. 2003. Electricity generation by direct oxidation of glucose in mediatorless microbial fuel cells. *Nat. Biotechnol.* 21:1229-1232.
29. Chen, F. & Dixon, R.A. 2007. Lignin modification improves fermentable sugar yields for biofuel production. *Nat. Biotechnol.* 25:759-761.
30. Cherubini, F. 2010. GHG balances of bioenergy systems - Overview of key steps in the production chain and methodological concerns. *Renew. Energ.* 35:1565–1573.
31. Chisti, Y. 2007. Biodiesel from micro*algae*. *Biotechnol. Adv.* 25:294-306.
32. Coleman, H.D., Park, J.-Y., Nair, R., Chapple, C. & Mansfield, S.D. 2008. RNAi-mediated suppression of p-coumaroyl-CoA

3′-hydroxylase in hybrid poplar impacts lignin deposition and soluble secondary metabolism. *Proc. Natl. Acad. Sci. USA* 105:4501- 4506.

33. Costanza, R. et al. 1997. The value of the world's ecosystem services and natural capital. *Nature* 387, 253-260.
34. Cropper, M., Geiger, S. & Jollie, D. 2004. Fuel cells: a survey of current developments. *J. Power Sources* 131:57–61.
35. Damartzis, T. & Zabaniotou, A. 2011. Thermochemical conversion of biomass to second generation biofuels through integrated process design-A review. *Renew. Sustain. Energ. Rev.* 15:366–378.
36. Darussalam, B. 2007. Last-Gasp effort to save Borneo's tropical rainforests. *Science* 317:192. Deluga, G.A., Salge, J.R., Schmidt, L.D. & Verykios, X.E. 2004. Renewable hydrogen from ethanol by autothermal reforming. *Science* 13:993-997.
37. Demirbas, M.F. & Balat, M. 2006. Recent advances on the production and utilization trends of bio-fuels: A global perspective. *Energ. Convers. Manage.* 47:2371-2381.
38. Demirbas, A. 2007. The influence of temperature on the yields of compounds existing in biooils obtaining from biomass samples via pyrolysis. *Fuel process. Technol.* 88:591-597. Demirbas, A. 2008a. Biofuels sources, biofuel policy, biofuel economy and global biofuel projections. *Energ. Convers. Manage.* 49:2106-2116.
39. Demirbas, A. 2008b. Comparison of transesterification methods for production of biodiesel from vegetable oils and fats. *Energ. Convers. Manage.* 49:125-130.
40. Demirbas A. 2010. Use of *algae* as biofuel sources. *Energ. Convers. Manage.* 51:2738–2749. de Vries, B.J.M., van Vuuren, D.P. & Hoogwijk, M.M. 2007. Renewable energy sources: Their global potential for the first-half of the 21st century at a global level: An integrated approach. *Energ. Policy* 35:2590-2610.
41. Dijkstra, A.J. 2006. Revisiting the formation of trans isomers during partial hydrogenation of triacylglycerol oils. *Eur. J. Lipid Sci. Tech.* 108:249-264.

42. Doniger, D.D., Herzog, A.V. & Lashof, D.A. 2006. An ambitious, centrist approach to global warming legislation. *Science* 314:764-765.

43. Donner, S.D. & Kucharik, C.J. 2008. Corn-based ethanol production compromises goal of reducing nitrogen export by the Mississippi River. *Proc. Natl. Acad. Sci.* USA 105:4513-4518.

44. Dorian, J.P., Franssen, H.T. & Simbeck, D.R. 2006. Global challenges in energy. *Energ. Policy* 34:1984-1991.

45. Easterling, E.R., French, W.T., Hernandez, R. & Licha, M. 2008. The effect of glycerol as a sole and secondary substrate on the growth and fatty acid composition of *Rhodotorula glutinis*. *Bioresource Technol.* 100:356-61

46. Eugenia Corria, M., Melian Cobas, V. & Silva Lora, E. 2006. Perspectives of Stirling engines use for distributed generation in Brazil. *Energ. Policy* 34:3402-3408.

47. Fagernäs, L., Brammer, J., Wilén, C., Lauer, M.& Verhoeff, F. 2010 Drying of biomass for second generation synfuel production. *Biomass Bioenergy* 34:1267-1277.

48. Fairless, D. 2007. The little shrub that could - maybe. *Nature* 449:652-655.

49. Fargione, J., Hill, J., Tilman, D., Polasky, S. & Hawthorne, P. 2008. Land clearing and the biofuel carbon debt. *Science* 319:1235-1238.

50. Fenning, T.M., Walter, C. & Gartland, K.M.A. 2008. Forest biotech and climate change. *Nat. Biotechnol.* 26:615-616.

51. Fialkov, A.B., Gordin, A. & Amirav, A. 2008. Hydrocarbons and fuels analyses with the supersonic gas chromatography mass spectrometry - The novel concept of isomer abundance analysis. *J. Chromatogr.* A 1195:127-135.

52. Fischer, C.R., Klein-Marcuschamer, D. & Stephanopoulos, G. 2008. Selection and optimization of microbial hosts for biofuel production. *Metab. Eng.* 10:295–304.

53. Fischer, G., Prieler, S., van Velthuizen, H., Lensink, S.M., Londo, M. & de Wit, M. 2010. Biofuel production potentials in Europe: Sustainable use of cultivated land and pastures. Part I: Land productivity potentials. *Biomass Bioenergy* 34:159 - 172.

54. Flavin, C. 2006. American Energy: The Renewable Path to Energy Security. 40 p. Washington DC: Worldwatch Institute.

55. Folkesson, A., Andersson, C., Alvfors, P., Alaküla, M. & Overgaard, L. 2003. Real life testing of a Hybrid PEM Fuel Cell Bus. *J. Power Sources* 118:349-357.

56. Fredrickson, J.K. et al. 2008. Towards environmental systems biology of *Shewanella*. *Nat. Rev. Microbiol.* 6:592-603.

57. Frondel, M. & Peters, J. 2007. Biodiesel: a new Oildorado? *Energ. Policy* 35:1675-84. García-Montero, L.G., López, E., Monzón, A. & Pastor, I.O. 2010. Environmental screening tools for assessment of infrastructure plans based on biodiversity preservation and global warming (PEIT, Spain). *Environ. Impact Assess. Rev.* 30:58–168.

58. Gavrilescu, M. & Chisti, Y. 2005. Biotechnology - a sustainable alternative for chemical industry. *Biotechnol. Adv.* 23:471-499.

59. Gielen, D.J., Fujino, J., Hashimoto, S. & Moriguchi, Y. 2002. Biomass strategies for climate policies? *Climate Policy* 2:319–333.

60. Glibert, P.M. et al. 2008. Ocean urea fertilization for carbon credits poses high ecological risks. *Mar. Pollut. Bull.* 56:1049-1056.

61. Gnansounou, E., Dauriat, A., Villegas, J. & Panichelli, L. 2009. Life cycle assessment of biofuels: Energy and greenhouse gas balances. *Bioresource Technol.* 100:4919–4930. Goering, E., Schwab, W., Daugherty, J., Pryde, H. & Heakin, J. 1982. Fuel properties of eleven vegetable oils. *Transactions ASABE* 25:1472-1483.

62. Goldemberg, J., Coelho, S.T. & Lucon, O. 2004. How adequate policies can push renewables. *Energ. Policy* 32:1141-1146.

63. Gressel, J. 2008. Transgenics are imperative for biofuel crops. *Plant Sci.* 174:246-263. Greyvenstein, R., Correia, M. & Kriel, W. 2008. South Africa's opportunity to maximise the role of nuclear power in a global hydrogen economy. *Nucl. Eng. Des.* 238:3031-3040.

64. Gui, M.M., Lee, K.T. & Bhatia, S. 2008. Feasibility of edible oil vs. non-edible oil vs. waste edible oil as biodiesel feedstock. *Energy* 33:1646-1653.

65. Gummer, J.H. & Head, C.R. 2003. Hydrogen, hydropower and world poverty. *Int. J. Hydropower Dams* 5:67-71. Haag, A.L. 2007. *Algae* bloom again. *Nature* 447:520-521. Hagerman, S., Dowlatabadi, H., Satterfield, T. & McDaniels, T. 2010. Expert views on biodiversity conservation in an era of climate change. *Global Environ. Change* 20:192–207. Hall, J., Matos, S., Severino, L. & Beltrão, N. 2009. Brazilian biofuels and social exclusion: established and concentrated ethanol versus emerging and dispersed biodiesel. *J. Clean. Prod.* 17:S77–S85.

66. Haug, C., Huitema, D. & Wenzler, I. 2011. Learning through games? Evaluating the learning effect of a policy exercise on European climate policy. *Technol. Forecast. Soc. Change*, doi:10.1016/j.techfore.2010.12.001

67. Havlík, P. et al. 2010. Global land-use implications of first and second generation biofuel targets. *Energ. Policy*, doi:10.1016/j.enpol.2010.03.030.

68. Hawken, P., Lovins, A. & Lovins, L.H. 2002. Natural Capitalism: Creating the Next Industrial Revolution. London: Earthscan. Accessed February 2, 2010. http://www.natcap.org/. Herrera, S. 2006. Bonkers about biofuels. *Nat. Biotechnol.* 24:755-760.

69. Heyer, N.I. & Woodward, J. 2001. Efficient hydrogen production using enzymes of the pentose phosphate cycle. Proceedings of the 2001 DOE Hydrogen Program Review, NREL/CP-570-30535. Accessed February 2, 2010. http://www1.eere.energy.gov/hydrogenandfuelcells/pdfs/30535g.pdf.

70. Hill, J., Nelson, E., Tilman, D., Polasky, S. & Tiffany D. 2006. Environmental, economic, and energetic costs and benefits of biodiesel and ethanol biofuels. *Proc. Natl. Acad. Sci. USA* 103:11206-11210.

71. Himmel, M.E., Ding, S.-Y., Johnson, D.K., Adney, W.S., Nimlos, M.R., Brady, J.W. & Foust, T.D. 2007. Biomass Recalcitrance: Engineering Plants and Enzymes for Biofuels Production. *Science* 315:804-807. Hissel, D., Péra, M.C. & Kauffmann, J.M. 2004. Diagnosis of automotive fuel cell power generators. *J. Power Sources* 128:239-246.

72. Hopkin, M. 2007. Climate sceptics switch focus to economics. *Nature* 445:582-583. Howden, S.M., Soussana, J.-F., Tubiello, F.N., Chhetri, N., Dunlop, M. & Meinke, H. 2007. Adapting

agriculture to climate change. *Proc. Natl. Acad. Sci. USA* 104:19691-19696. Huston, M.A. & Marland, G. 2003. Carbon management and biodiversity. *J. Environ. Manage.* 67:77-86.

73. Johnston, M. & Holloway, T. 2007. A global comparison of national biodiesel production potentials. *Environ. Sci. Technol* 41:7967-7973.
74. Johnston, C.D., Bastow, T.P. & Innes, N.L. 2007. The use of biodegradation signatures and biomarkers to differentiate spills of petroleum hydrocarbon liquids in the subsurface and estimate natural mass loss. *Eur. J. Soil Biol.* 43:328-334.
75. Jones, D.B. 2010. Proposal for an international carbon fee. *Energy for Sustainable Development* 14:245–250.
76. Kalscheuer, R., Stoveken, T. & Steinbuchel, A. 2006a. Microdiesel: *Escherichia coli* engineered for fuel production. *Microbiology* 152:2529-2536.
77. Kalscheuer, R., Stoveken, T., Luftmann, H., Malkus, U., Reichelt, R. & Steinbuchel, A. 2006b. Neutral lipid biosynthesis in engineered *Escherichia coli*: Jojoba oil-like wax esters and fatty acid butyl esters. *Appl. Environ. Microbiol.* 72:1373-1379.
78. Kaplan, I.R., Galperin, Y., Lu, S.-T. & Lee, R.-P. 1997. Forensic environmental geochemistry: differenciation of fuel-types, their sources and release time. *Org. Geochem.* 27:289- 317. Khaselev, O. & Turner, J. 1998. A monolithic photovolatic–photochemical device for hydrogen production via water splitting. *Science* 280, 425-427.
79. Kikuchi, R. 2006. Penetration of hydrogen-based energy system and its potential for causing global environmental change: Scoping risk analysis based on life cycle thinking. *Environ. Impact Asses. Rev.* 26:206-218.
80. Kindermann, G., Obersteiner, M., Sohngen, B., Sathaye, J., Andrasko, K., Rametsteiner, E., Schlamadinger, B., Wunder, S. & Beach, R. 2008. Global cost estimates of reducing carbon emissions through avoided deforestation. *Proc. Natl. Acad. Sci. USA* 105:10302-10307. King, C.W., Holman, A.S. & Webber, M.E. 2008. Thirst for energy. *Nat. Geosci.* 1:283-286.
81. Kirilenko, A.P. & Sedjo, R.A. 2007. Climate change impacts on forestry. *Proc. Natl. Acad. Sci. USA* 104:19697-19702.

82. Kleiner, K. 2007. Civil aviation faces green challenge. *Nature* 448:120-121.

83. Kleiner, K. 2008. The backlash against biofuels. *Nat. Rep. Clim. Change* 2:9-11, doi:10.1038/climate.2007.71.

84. Kremer, F.G. & Fachetti, A. 2000 Alcohol as automotive fuel-Brazilian experience. SAE Technical Paper No. 2000-01-1965, doi:10.4271/2000-01-1965

85. Kumar, A. & Sharma, S. 2008. An evaluation of multipurpose oil seed crop for industrial uses (*Jatropha curcas* L.): A review. *Ind. Crop. Prod.* 28:1-10.

86. Lamy, C., Rousseau, S., Belgsir, E.M., Coutanceau, C. & Léger, JM. 2004. Recent progress in the direct ethanol fuel cell: development of new platinum–tin electrocatalysts. *Electrochim. Acta* 49:3901-3908.

87. Laurance, W.F. 2007. Switch to corn promotes amazon deforestation. *Science* 318:1721. Leibold, H., Hornung, A. & Seifert, H. 2008. HTHP syngas cleaning concept of two stage biomass gasification for FT synthesis. *Powder Technol.* 180:265-270.

88. León-Bañares, R., González-Ballester, D., Galván, A. & Fernández, E. 2004. Transgenic micro*algae* as green cellfactories. *Trends Biotechnol.* 22:45-52.

89. Lewis, N.S. & Nocera, D.G. 2006. Powering the planet: Chemical challenges in solar energy utilization. *Proc. Natl. Acad. Sci. USA* 103:15729-15735.

90. Lewis, N.S. 2007. Toward cost-effective solar energy use. *Science* 315:798-801.

91. Lima, D.G. et al. 2004. Diesel-like fuel obtained by pyrolysis of vegetable oils. *J. Anal. Appl. Pyrol.* 71:987-996.

92. Lin, C.-Y. et al. 2010. A pilot-scale high-rate biohydrogen production system with mixed microflora. *Int. J. Hydrogen Energ.*, doi:10.1016/j.ijhydene.2010.07.115.

93. Liu, X., Li, X., Suehiro, Y. & Fujimoto, K. (2007). Elucidation of reaction network and effective control of carbon number distribution in the three phase Fischer-Tropsch synthesis. *Appl. Catal. A-Gen.* 333:211-218.

94. Lois, E. 2007. Definition of biodiesel. *Fuel* 86:1212-1213.

95. Lokey, E. 2007. How the next US president should slow global warming. *Energ. Policy* 35:5399-5402.

96. Lonngren, K.E. & Bai, E.-W. 2008. On the global warming problem due to carbon dioxide. *Energ. Policy* 36:1567-1568.

97. Lovelock, J.E. & Margulis, L. 1974. Atmospheric homeostasis by and for the biosphere- The Gaia hypothesis. *Tellus* 26:2-10.

98. Lü, X-c, L., Yang, J-g, Zhang, W-g, Z. & Huang, Z. 2004. Effect of cetane number improver on heat release rate and emissions of high speed diesel engine fueled with ethanol– diesel blend fuel. *Fuel* 83:2013-2020.

99. Luo, T.-J.M., Soong, R., Lan, E., Dunn, B. & Montemagno, C. 2005. Photo-induced proton gradients and ATP biosynthesis produced by vesicles encapsulated in a silica matrix. *Nat. Materials* 4:220–224.

100. Malça, J. & Freire, F. 2011. Life-cycle studies of biodiesel in Europe: A review addressing the variability of results and modeling issues. *Renew. Sustain. Energ. Rev.* 15:338-351. Mäler, K.-G., Aniyar, S. & Jansson, Å. 2008. Accounting for ecosystem services as a way to understand the requirements for sustainable development. *Proc. Natl. Acad. Sci. USA* 105:9501-9506.

101. Malhi, Y., Roberts, J.T., Betts, R.A., Killeen, T.J., Li, W. & Nobre, C.A. 2008. Climate change, deforestation, and the fate of the Amazon. *Science* 319:169-172.

102. Mander, U. et al. 2010. Assessment of methane and nitrous oxide fluxes in rural landscapes. *Landscape Urban Plan.* 98:172–181.

103. Marchetti, J., Miguel, V. & Errazu, A. 2007. Possible methods for biodiesel production. *Renew. Sustain. Energ. Rev.* 11:1300-1311.

104. Marcinkoski, J., Kopasz, J.P. & Benjamin, T.G. 2008. Progress in the US DOE fuel cell subprogram efforts in polymer electrolyte fuel cells. *Int. J. Hydrogen Energ.* 33:3894- 3902. Markels, M. & Barber, R.T. 2001. Sequestration of carbon dioxide by ocean fertilization. Preprints - American Chemical Society, Division of Petroleum Chemistry 46:45-48.

105. Markou, G. & Georgakakis, D. 2011. Cultivation of filamentous cyanobacteria (blue-green *algae*) in agro-industrial wastes and wastewaters: A review. *Appl. Energ.*, doi:10.1016/j.apenergy.2010.12.042

106. Marris, E. 2006. Drink the best and drive the rest. *Nature* 444:670-672.

107. Mathews, J. & Wang, G. 2009. Metabolic pathway engineering for enhanced biohydrogen production. *Int. J. Hydrogen Energ.* 34:7404 – 7416.

108. Meng, X., Yang, J., Xu, X., Zhang, L., Nie, Q. & Xian, M. 2008. Biodiesel production from oleaginous microorganisms. *Renew. Energ.* 34:1-5

109. Michel, D. 2009. Foxes, hedgehogs, and greenhouse governance: Knowledge, uncertainty, and international policy-making in a warming World. *Appl. Energ.* 86:258–264.

110. Miles, L. & Kapos, V. 2008. Reducing greenhouse gas emissions from deforestation and forest degradation: global land-use implications. *Science* 320:1454-1455.

111. Mills, G. et al. 2010. climate information for improved planning and management of mega cities (needs perspective). *Procedia Environmental Sciences* 1:228–246

112. Moreira, J.R. & Goldemberg, J. 1999. The alcohol program. *Energ. Policy* 27:229-245. Moriarty, P. & Honnery, D. 2007. The prospects for global green car mobility. *J. Clean. Prod.* 16:1717-1726.

113. Nadeau, R. 2006. The Environment Endgame: Mainstream Economics, Ecological Disaster, and Human Survival. 214 p. London: Rutgers University Press.

114. Nass, L.L., Pereira, P.A.A. & Ellis, D. 2007. Biofuels in Brazil: An Overview. *Crop Sci.* 47:2228-2237.

115. Nast, M., Langniss, O. & Leprich, U. 2007. Instruments to promote renewable energy in the German heat market-Renewable Heat Sources Act. *Renew. Energ.* 32:1127–1135.

116. Odling-Smee, L. 2007. Biofuels bandwagon hits a rut. *Nature* 446:483.

117. Okereke, C. & Dooley, K. 2010. Principles of justice in proposals and policy approaches to avoided deforestation: Towards a post-Kyoto climate agreement. *Global Environ. Change* 20:82–95.

118. Om Tapanes, N.C., Gomes Aranda, D.A., de Mesquita Carneiro, J.W. & Ceva Antunes, O.A. 2008. Transesterification of *Jatropha*

curcas oil glycerides: Theoretical and experimental studies of biodiesel reaction. *Fuel* 87:2286-2295.

119. Orellana, C. & Neto, R.B. 2006. Brazil and Japan give fuel to ethanol market. *Nat. Biotechnol*. 24:232.
120. Parker, J. 2002. Turning manure into gold. *EMBO rep*. 3:1114-1116.
121. Parry, M., Palutikof, J., Hanson, C. & Lowe, J. 2008. Squaring up to reality. *Nature Reports Climate Change* 2, 68-70.
122. Peters, W. et al. 2007. An atmospheric perspective on North American carbon dioxide exchange: CarbonTracker. *Proc. Natl. Acad. Sci. USA* 104:18925-18930.
123. Pousa, G.P.A.G., Santos, A.L.F. & Suarez, P.A.Z. 2007. History and policy of biodiesel in Brazil. *Energ. Policy* 35:5393-5398.
124. Proctor, M.R., Taylor, E.J., Nurizzo, D., Turkenburg, J.P., Lloyd, R.M., Vardakou, M., Davies, G.J. & Gilbert, H.J. 2005. Tailored catalysts for plant cell-wall degradation: Redesigning the exo/endo preference of *Cellvibrio japonicus* arabinanase 43A. *Proc. Natl. Acad. Sci. USA* 102:2697-2702.
125. Pryor, S.C. & Barthelmie, R.J. 2010. Climate change impacts on wind energy: A review. Renew. Sustain. *Energ. Rev*. 14:430–437.
126. Pulz, O. & Gross, W. 2004. Valuable products from biotechnology of micro*algae*. *Appl. Microbiol. Biot*. 65:635-48.
127. Rafaj, P. & Kypreos, S. 2007. Internalisation of external cost in the power generation sector: Analysis with global multi-regional MARKAL model. *Energ. Policy* 35:828-843.
128. Ragauskas, A.J. et al. 2006. The path forward for biofuels and biomaterials. *Science* 311:484- 489.
129. Raghu, S., Anderson, R.C., Daehler, C.C., Davis, A.S., Wiedenmann, R.N., Simberloff, D. & Mack, R.N. 2006. Adding biofuels to the invasive species fire? *Science* 313:1742.
130. Ramadhas, A.S., Jayaraj, S. & Muraleedharan, C. 2005. Biodiesel production from high FFA rubber seed oil. *Fuel* 84:335-340.
131. Ratledge, C. 2002. Regulation of lipid accumulation in oleaginous micro-organisms. *Biochem. Soc. Trans*. 30:1047-1050.
132. Reay, D. 2007. Spring-time for sinks. *Nature* 446:727-728.

133. Ridgwell, A.T. & Zeebe, R.E. 2005. The role of the global carbonate cycle in the regulation and evolution of the Earth system. *Earth Planet. Sci. Lett.* 234:299-315.

134. Romm, J. 2006. The car and fuel of the future. *Energ. Policy* 34:2609-2614.

135. Ross, D.K. 2006. Hydrogen storage: The major technological barrier to the development of hydrogen fuel cell cars. *Vacuum* 80:1084-1089.

136. Rothengatter, W. 2010. Climate change and the contribution of transport: Basic facts and the role of aviation. *Transportation Research* Part D 15:5–13.

137. Rout, U.K., Akimoto, K., Sano, F., Oda, J., Homma, T. & Tomoda, T. 2008. Impact assessment of the increase in fossil fuel prices on the global energy system, with and without CO_2 concentration stabilization. *Energ. Policy* 36:3477-3484.

138. Sánchez Mirón, A. Ceron Garcia, M.C., Contreras Gomez, A., Garcia Camacho, F., Molina Grima, E. & Chisti, Y. 2003. Shear stress tolerance and biochemical characterization of *Phaeodactylum tricornutum* in quasi steady-state continuous culture in outdoor photobioreactors. *Biochem. Eng. J.* 16:287-297.

139. Sanderson, K. 2006. A field in ferment. *Nature* 444:673-676.

140. Scarlat, N. & Dallemand, J.-F. 2010. Recent developments of biofuels/bioenergy sustainability certification: A global overview. *Energ. Policy*, doi:10.1016/ j.enpol.2010.12.039

141. Scharlemann, J.P.W. & Laurance, W.F. 2008. How green are biofuels? *Science* 319:43-44. Schiermeier, Q. 2008. Europe spells out action plan for emissions targets. *Nature* 451:504-505.

142. Schiermeier, Q., Tollefson, J., Scully, T., Witze, A. & Morton, O. 2008. Electricity without carbon. *Nature* 454:816-823.

143. Schmer, M.R., Vogel, K.P., Mitchell, R.B. & Perrin, R.K. 2008. Net energy of cellulosic ethanol from switchgrass. *Proc. Natl. Acad. Sci. USA* 105:464-469.

144. Schultz, M.G., Diehl, T., Brasseur, G.P. & Zittel, W. 2003. Air pollution and climate-forcing impacts of a global hydrogen economy. *Science* 302:624-627.

145. Searchinger, T. et al. 2008. Use of U.S. croplands for biofuels increases greenhouse gases through emissions from land-use change. *Science* 319 :1238-1240.

146. Sedjo, R.A. & Marland, G. 2003. Inter-trading permanent emissions credits and rented temporary carbon emissions offsets: some issues and alternatives. *Climate Policy* 3:435-444.

147. Seibert, M., Forestier, M., Zhang, L. & Ghirardi, M. 2001. Development of an efficient algal H2-production system. Proceedings of the 2001 DOE Hydrogen Program Review NREL/CP-570-30535, Accessed February 2, 2010. http://www1.eere.energy.gov/hydrogenandfuelcells/pdfs/30535c.pdf.

148. Shinnar, R. & Citro, F. 2006. A road map to U.S. decarbonization. *Science* 313:1243-1244. Simões, A.F., Kligerman, D.C., La Rovere, E.L., Maroun, M.R., Barata, M. & Obermaie, M. 2010. Enhancing adaptive capacity to climate change: The case of smallholder farmers in the Brazilian semi-arid region. *Environmental Science & Policy* 13:801-808.

149. Singh, J. & Gu, S. 2010. Commercialization potential of micro*algae* for biofuels production. *Renew. Sustain. Energ. Rev.* 14:2596–2610.

150. Snyder, C.S., Bruulsema, T.W., Jensen, T.L. & Fixen, P.E. 2009 Review of greenhouse gas emissions from crop production systems and fertilizer management effects. *Agric. Ecosyst. Environ.* 133:247–266.

151. Solomon, B.D. & Banerjee, A. 2006. A global survey of hydrogen energy research, development and policy. *Energ. Policy* 34:781-792.

152. Song, C. 2006. Global challenges and strategies for control, conversion and utilization of CO2 for sustainable development involving energy, catalysis, adsorption and chemical processing. *Catalysis Today* 115:2-32.

153. Song, C.S. & Pan, W. 2004. Tri-reforming of methane: a novel concept for catalytic production of industrially useful synthesis gas with desired H/CO ratios. *Catalysis Today* 98:463-484.

154. Song, D., Fu, J. & Shi, D. 2008. Exploitation of oil-bearing micro*algae* for biodiesel. *Chinese Journal of Biotechnology* 24:341-348.

155. Soong, R.K., Bachand, G.D., Neves, H.P., Olkhovets, A.G., Craighead, H.G. & Montemagno, C.D. 2000. Powering an inorganic nanodevice with a biomolecular motor. *Science* 290:1555-1558.

156. Sorensen, B. 2007. Assessing current vehicle performance and simulating the performance of hydrogen and hybrid cars. *Int. J. Hydrogen Energ.* 32:1597-1604.

157. Spolaore, P., Joannis-Cassan, C., Duran, E., & Isambert, A. (2006). Commercial applications of micro*algae. J. Biosci. Bioeng.* 101:87-96.

158. Steinberg, M. 2006. Conversion of fossil and biomass fuels to electric power and transportation fuels by high efficiency integrated plasma fuel cell (IPFC) energy cycle. *Int. J. Hydrogen Energ.* 31:405-411.

159. Stern, N. 2006. Review on the economics of climate change. Accessed February 2, 2010. http://www.hmtreasury. gov.uk/ independent_reviews/stern_review_economics_climate_ change/ stern_review_report.cfm.

160. Sticklen, M.B. 2008. Plant genetic engineering for biofuel production: towards affordable cellulosic ethanol. *Nature Rev. Genet.* 9:433-443.

161. Stone, R. 2007. Can palm oil plantations come clean? *Science* 317:1491.

162. Styles, D. & Jones, M.B. 2008. Life-cycle environmental and economic impacts of energycrop fuel-chains: an integrated assessment of potential GHG avoidance in Ireland. *Environmental Science and Policy* 11:294-306.

163. Subhadra, BG. 2010 Sustainability of algal biofuel production using integrated renewable energy park (IREP) and algal biorefinery approach. *Energ. Policy* 38:5892–5901.

164. Subramanian, K,A., Singal, S.K., Saxena, M. & Singhal, S. 2005. Utilization of liquid biofuels in automotive Diesel engines: an Indian perspective. *Biomass Bioenergy* 29:65-72.

165. Sun, F. & Chen, H. 2008. Enhanced enzymatic hydrolysis of wheat straw by aqueous glycerol pretreatment. *Bioresource Technol.* 99:6156-6161.

166. Szulczyk, K.R., McCarl, B.A. & Cornforth, G. 2010. Market penetration of ethanol. *Renew. Sustain. Energ. Rev.* 14:394–403.

167. Takeshita, T. & Yamaji, K. 2008. Important roles of Fischer-Tropsch synfuels in the global energy future. *Energ. Policy* 36:2773-2784.

168. Tan, K.T., Lee, K.T. & Mohamed, A.R. 2008. Role of energy policy in renewable energy accomplishment: The case of second-generation bioethanol. *Energ. Policy* 36:3360- 3365.

169. Theis, M., Skrifvars, B.-J., Zevenhoven, M., Hupa, M. & Tran, H. 2006. Fouling tendency of ash resulting from burning mixtures of biofuels. Part 2: Deposit chemistry. *Fuel* 85:1992-2001.

170. Thomas, C.D. 2007. A sixth mass extinction? *Nature* 450:349.

171. Tijmensen, M.J.A., Faaij, A.P.C., Hamelinck, C.N. & Van Hardeveld, M.R.M. 2002. Exploration of the possibilities for production of Fischer Tropsch liquids and power via biomass gasification. *Biomass Bioenergy* 23:129-152.

172. Tilman, D., Hill, J. & Lehman, C. 2007. Response to comment on "carbon-negative biofuels from low-input high-diversity grassland biomass". *Science* 316:1567c.

173. Tollefson, J. 2008. Stoking the fire. *Nature* 454:388-392.

174. Tomkiewicz, M. 2006. Global warming: science, money and self-preservation. *Comptes rendus Chimie* 9 :172-179.

175. Trabesinger, A. 2007. Power games. *Nature* 447:900-903.

176. UNFCCC 2002. Report of the Conference of the Parties on its Seventh Session, Marrakesh, 29 October to 10 November 2001, Document FCCC/CP/2001/13/add.1

177. Urry, J. 2008. Governance, flows, and the end of the car system? *Global Environ. Change* 18:343-349.

178. Velu, S., Suzuki, K., Vijayaraj, M., Barman, S. & Gopinath, C. 2005. In situ XPS investigations of Cu1_xNixZnAl-mixed metal oxide catalysts used in the oxidative steam reforming of bio-ethanol. *Applied Catalysis* B: Environmental 55:287-299.

179. Venter, O., Meijaard, E. & Wilson, K. 2008. Strategies and alliances needed to protect forest from palm-oil industry. *Nature* 451:16.

180. Venturi, P. & Venturi, G. 2003. Analysis of energy comparison for crops in European agricultural systems. *Biomass Bioenergy* 25:235-255.

181. Vertès, A.A., Inui, M. & Yukawa, H. 2006. Implementing biofuels on a global scale. *Nat. Biotechnol.* 24:761-764.

182. Vogt, K.A. et al. 2008. Bio-methanol: How energy choices in the western United States can help mitigate global climate change. *Renew. Energ.* 34:233-241.

183. Vunjak-Novakovic, D., Kim, Y., Wu, X., Berzin, I. & Merchuk, J.C. 2005. Air-lift bioreactors for algal growth on flue gas: mathematical modeling and pilot-plant studies. *Ind. Eng. Chem. Res.* 44:6154-6163.

184. Wall, D. 2004. Rain or shine – a phototroph that delivers. *Nat. Biotechnol.* 22:40-41.

185. Walter, A., Dolzan, P., Quilodrán, O., de Oliveira, J.G., da Silva, C., Piacente, F. & Segerstedt, A. 2010. Sustainability assessment of bio-ethanol production in Brazil considering land use change, GHG emissions and socio-economic aspects. *Energ. Policy*, doi:10.1016/j.enpol.2010.07.043

186. Werner, F., Taverna, R., Hofer, P., Thürig, E. & Kaufmann, E. 2010. National and global greenhouse gas dynamics of different forest management and wood use scenarios: a model-based assessment. *Environmental Science & policy* 13:72 – 85.

187. Wesseler, J. 2007. Opportunities ('costs) matter: A comment on Pimentel and Patzek "Ethanol production using corn, switchgrass, and wood; biodiesel production using soybean and sunflower". *Energ. Policy* 35:1414-1416.

188. Willner, I. 2002. Biomaterials for sensors, fuel cells, and circuitry. *Science* 298:2407-2408. Wolfrum, E.J. & Watt, A.S. 2001. Bioreactor design studies for a novel hydrogen-producing bacterium. Proceedings of the 2001 DOE Hydrogen Program Review, NREL/CP- 570-30535.

189. Woodcock, J., Banister, D., Edwards, P., Prentice, A.M. & Roberts, I. 2007. Energy and health 3, energy and transport. *Lancet* 370:1078-1088.

190. Xue, F., Miao, J., Zhang, X., Luo, H. & Tan, T. 2008. Studies on lipid production by *Rhodotorula glutinis* fermentation

using monosodium glutamate wastewater as culture medium. *Bioresource Technol.* 99:5923-5927.

191. Yang, Z. & Sirianni, P. 2010. Balancing contemporary fairness and historical justice: A 'quasiequitable' proposal for GHG mitigations. *Energ. Econ.* 32:1121–1130.
192. Yasuda, R., Noji, H., Yoshida, M., Kinosita, K. & Itoh, H. 2001. Resolution of distinct rotational substeps by submillisecond kinetic analysis of F1-ATPase. *Nature* 410:898-904.
193. Yue, T.-X., Fan, Z.-M., Chen, C.-F., Sun, X.-F.& Li, B.-L. 2010. Surface modeling of global terrestrial ecosystems under three climate change scenarios. *Ecol. Model.*, doi:10.1016/j.ecolmodel.2010.11.026
194. Zhang, Y., Dubé, M.A., McLean, D.D. & Kates, M. 2003. Biodiesel production from waste cooking oil: Process design and technological assessment. *Bioresource Technol.* 89:1- 16.
195. Zhou, X., Xiao, B., Ochieng, R.M. & Yang, J. 2008. Utilization of carbon-negative biofuels from low-input high-diversity grassland biomass for energy in China. *Renew. Sustain. Energ. Rev.* 13:479-485.

Chapter 2

Biomass Feedstock Pre-Processing– Part 1: Pre-Treatment

Tail Lope[1], Adapt Phani[1], and Kashaninejad Mahdi[2]

[1] Department of Chemical and Biological Engineering, University of Saskatchewan, Canada

[2] Department of Food Science & Technology, Groan University of Agricultural Sciences and Natural Resources Groan, Iran

INTRODUCTION

Sources of Biomass

The two main sources of biomass for energy generation are purpose-grown energy crops and waste materials (Larkin et al., 2004). Energy crops, such as Miscanthus and short rotation woody crops (coppice), are cultivated mainly for energy purposes and are associated with the food vs. fuels debate, which is concerned with whether land should be used for fuel rather than food production. The use of residues from agriculture, such as barley, canola, oat and wheat straw, for energy generation circumvents the food vs. fuel dilemma and adds value to existing crops (Chico-Santamaria et al., 2009). In fact, these residues represent an abundant, inexpensive and readily available source of renewable lignocellulose biomass (Liu et al., 2005).

Current Issues Related to Biomass Utilization

The main problem with agricultural straw is its relatively low density in its original or baled forms. The bulk density of loose and standard baled straw is approximately 40 kg/m^3 and 100 kg/m^3, respectively, compared with the bulk density of unprocessed wood residue, which is approximately 250 kg/m^3(Demarrias, 2001; Tripathi et al., 1988). The relative low density of straw makes it more expensive to transport compared to wood and coal because a lower mass of straw can be transported per unit volume. Additionally, a larger storage area/volume is required for baled straw compared to wood chip. Densification into pellets increases the bulk density of biomass (McMullen et al., 2005; Obernberger and Thek,2004) and as a result, the net calorific content per unit volume is increased (Bhattacharya et al., 1989) and the storage, transport and handling of the material is easier and cheaper (Balatinecz, 1983; Bhattacharya et al., 1989; Kalian and Morey, 2006).

The quality of fuel pellet is usually assessed based on its density and durability. High bulk density increases storage and transport capacity of pellets (Adapa et al., 2007; Mani et al., 2003). Since feeding of boilers and gasifiers generally is volume-dependent, variations in bulk density should be avoided (Larsson et al., 2008). A bulk density of 650 kg/m^3 is stated as design value for wood pellet producers (Obernberger and Thek 2004). Low durability of pellets results in problems like disturbance within pellet feeding systems, dust emissions, and increased risk of fire and explosions during pellet handling and storage (Temmerman et al. 2006).

Densification of straw and determining the optimal parameters involved is an art in itself. The entire process involves securing of baled straw from agricultural fields, size reduction (chopping and grinding), application of pre-treatment (chemical, physic-chemical, and biological), determining the physical and frictional properties of straw grinds, lignocellulose characterization of straw, lab-scale and pilot-scale densification of grinds into pellets to determine the effect of various independent parameters on quality (density and durability), and energy analysis/ balance (Fig. 1). This chapter will only address the effect and need of lignocellulose characterization, pre-treatment and size reduction, and physical properties on densification of agricultural straw.

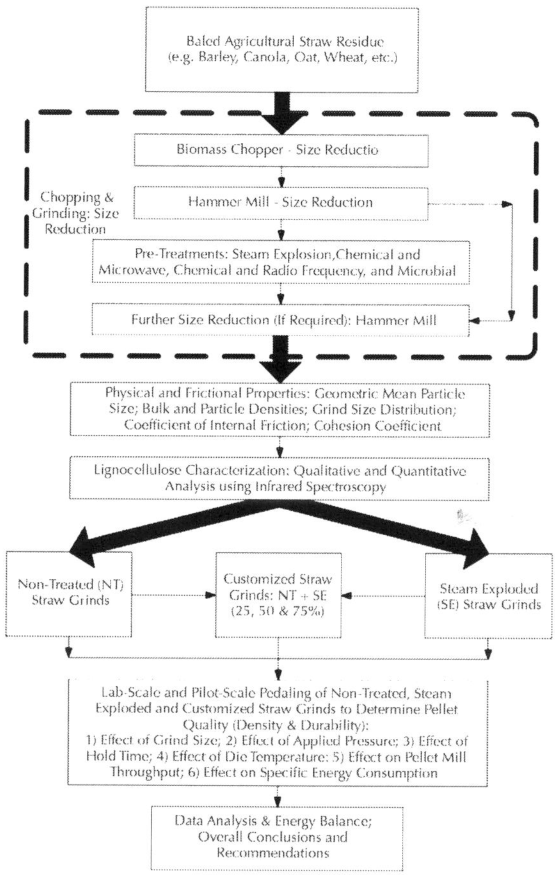

Figure 1: Processing steps involved in converting straw from field to pelletized product.

LIGNOCELLULOSIC BIOMASS CHARACTERIZATION

Structure of Lignocellulose Material

Lignocellulose material refers to plant biomass that is composed of cellulose, hemicellulose, and lignin (Fig. 2) (Lin and Tanaka, 2006). The major combustible component of non-food energy crops is cellulose, followed by lignin.

Cellulose: Cellulose is an organic polysaccharide consisting of a linear chain of several hundreds to over nine thousand (1→4) linked D-glucose $(C_6H_{10}O_5)$ n units (Crawford, 1981; Updegraff, 1969). Cellulose, a fibrous, tough, water-insoluble substance, is found in the cell walls of plants, particularly in the stalks, stems, trunks and all the woody portions of the plant body (Nelson and Cox, 2005). Cellulose comprises 40-60% of the dry weight of plant material (the cellulose content of cotton is 90% and that of wood is 50%) (Encyclopedia Britannica, 2008; USDE, 2006).

Zander son's et al. (2004) and Shaw (2008) reported that binding of wood material during hot pressing / densification is mainly dependent on the transition of cellulose into the amorphous state. According than (1989), due to the semi-crystalline structure, hydrogen bonded cellulose cannot be dissolved easily in conventional solvents, and it cannot be melted before it burns; hence, cellulose itself is not a suitable adhesive. This can be overcome by breaking the hydrogen bonds, thus making the cellulose molecule more flexible (Hon 1989). Cellulose requires a temperature of 320°C and pressure of 25 MPa to become amorphous in water (Douche et al., 2006).

Hemicellulose: Hemicellulose is made of several heteropolymers (matrix polysaccharides) present in almost all plant cell walls along with cellulose (Fig. 2). While cellulose is crystalline, strong, and resistant to hydrolysis; hemicellulose has a random, amorphous structure with less strength. Hemicellulose is a polysaccharide related to cellulose and comprises 20-40% of the biomass of most plants. In contrast to cellulose, hemicellulose is derived from several sugars in addition to glucose, including especially xylose but also mannose, GA lactose, ramose and arabinose (Shamble and Kennedy, 1985). Branching in hemicellulose produces an amorphous structure that is more easily hydrolyzed than cellulose (Shaw, 2008). Also, hemicellulose can be dissolved in strong alkali solutions. Hemicellulose provides structural integrity to the cell. Some researchers believe that natural bonding may occur due to the adhesive properties of degraded hemicellulose (Bhattacharya et al., 1989).

Lignin: Lignin is a complex chemical compound most commonly derived from wood and is an integral part of the cell walls of plants (Lebo et al., 2001; Zandersons et al., 2004). The compound has several unusual properties as a biopolymer, not the least its heterogeneity in lacking a defined primary structure. Lignin fills the spaces in the cell wall between cellulose and hemicellulose (Fig. 2). It is covalently linked to hemicellulose and thereby crosslinks different plant polysaccharides, conferring mechanical strength to the cell wall and consequently to the whole plant structure (Chabannes et al., 2001).

Lignin acts as a binder for the cellulose fibres (Fig. 2). van Dam et al. (2004) have reported that lignin can be used as an intrinsic resin in binderless board production due to the fact that when lignin melts (temperatures above 140°C), it exhibits thermosetting properties. Lignin is the component that permits adhesion in the wood structure, and is a rigidifying and bulking agent (Anglès et al., 2001). Lehtikangas (2001) reported that water (8-15%) in pellets will reduce the softening temperature of lignin to 100-135°C by plasticizing the molecular chains. The adhesive properties of thermally softened lignin are thought to contribute considerably to the strength characteristics of briquettes made of lignocellulosic materials (Granada et al., 2002; Shaw, 2008).

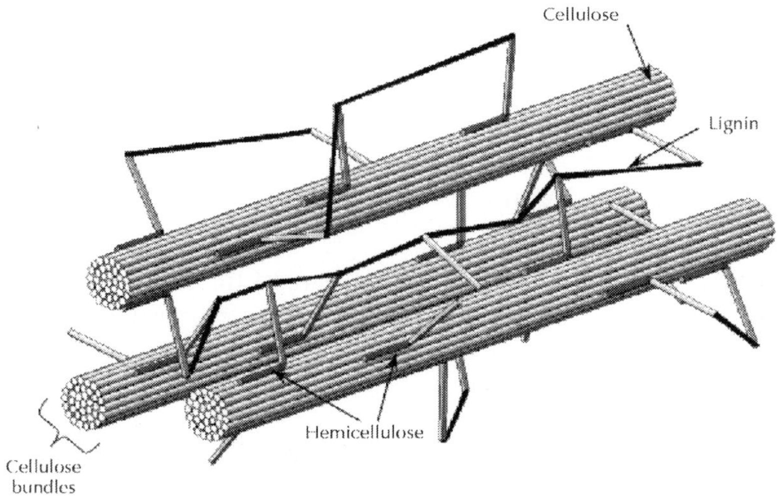

Figure 2: Location and arrangement of cellulose microfibrils in plant cell walls (Murphy and McCarthy, 2005; Shaw, 2008).

Rapid Characterization of Lignocellulosic Materials

The effect of various pre-processing and pre-treatment methods (Fig. 1) on the lignocellulosic matrix at the molecular level is not well understood. Applications of pre-processing methods such as size reduction or increasing porosity, and pre-treatment techniques such as steam explosion on agricultural biomass have demonstrated an improvement in pellet (compact) quality that can be attributed to the changes in the lignocellulosic components and distribution (Bagby, 1982; Focher et al., 1998). Therefore, it is critical to rapidly quantify the change in cellulose, hemicelluloses and lignin components of biomass due to application of pre-treatment methods.

Infrared spectroscopy has the potential to produce qualitative and quantitative analytical data for samples with minimum or no sample preparation, and at high speed and throughput (Adapa et al., 2011b and 2009; Budevska, 2002; Luypaert et al., 2003; Smola and Urleb, 2000; Tucker et al., 2000). Traditionally, chemical analyses of the individual components (e.g., lignin) of lignocellulosics have been performed by acid hydrolysis followed by gravimetric determination of lignin (Kelley et al., 2004). These methods can provide highly precise data. However, these methods are laborious, time-consuming, and, consequently, expensive to perform and sample throughput is limited.

Fourier Transform Infrared Spectroscopy

Fourier Transform Infrared Spectroscopy (FTIR) can be used to rapidly characterize and quantify cellulose-hemicellulose-lignin composition prior to and after application of various methods of pre-processing and pre-treatment of biomass (Adapa et al., 2009). The quantitative analysis of FTIR absorption spectrometry is based on the Bouguer-Beer-Lambert law (Sherman Hsu, 1997). According to this law, the intensities of absorption bands are linearly proportional to the concentration of each component in a homogenous mixture or solution.

Regression equations to predict the lignocellulosic content of agricultural biomass can be developed using pure cellulose, hemicelluloses and lignin as reference samples, and subsequently mixing them in different proportions to determine the change in absorption intensity at characteristic peak height (Adapa et al., 2011b). An overview of the experimental procedure to characterize the lignocellulosic composition is provided in Figure 3.

Pure cellulose has five distinct characteristic/ prominent peaks at wavenumbers of 1431, 1373, 1338, 1319 and 1203 cm^{-1}. Similarly, hemicellulose (xylan) has prominent peaks at wavenumbers of 1606, 1461, 1251, 1213, 1166 and 1050 cm^{-1}. The lignin spectrum has characteristic peaks at wavenumber of 1599, 1511, 1467, 1429, 1157 and 1054 cm^{-1}. The intensity of absorption at characteristic peak heights of cellulose, hemicellulose and lignin were used to develop regression equations to predict lignocellulosic composition of any agricultural biomass (Table 1) (Adapa et al., 2011b).

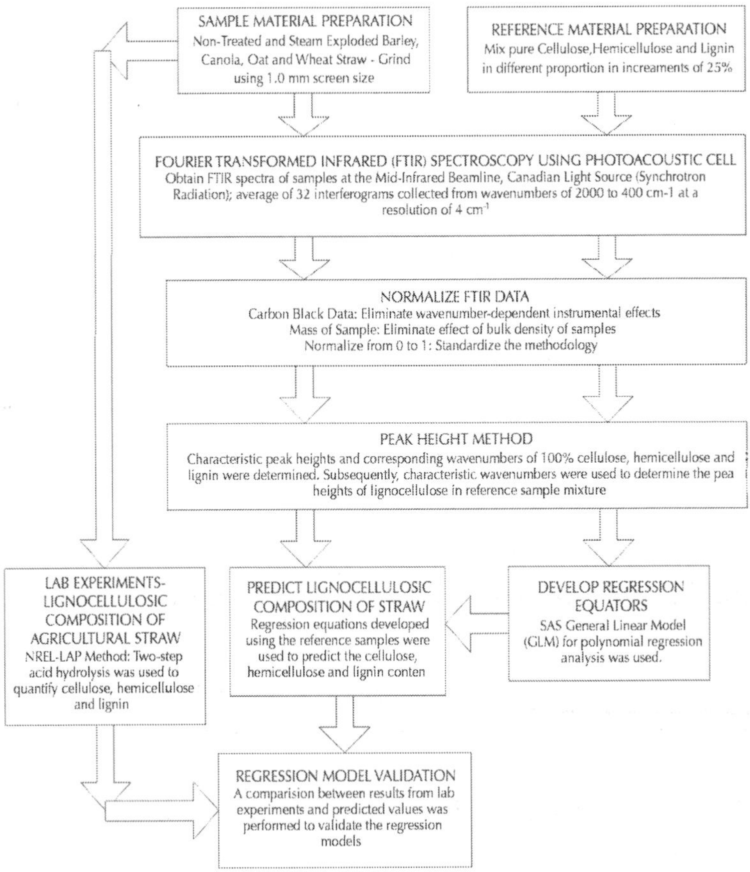

Figure 3: Experimental procedure followed to characterize lignocellulosic composition of agricultural straw (Adapa et al., 2011b).

Table 1: Regression equations to predict the lignocellulosic composition of agricultural biomass (Adapa et al., 2011b)

Equation	**% Mean Absolute Deviation**
%Cellulose=−135.10+781.35(PH_1319)−795.57(PH_1431)−135.26(PH_1203)+436.11(PH_1338)−94.24(PH_1373)	7.5

%Hemicellulose=1638.72−2581.71(PH_1251×PH_1461)−1260.90(PH_1213)−2518.05(PH_1166)+1573.69(PH_1213×PH_1251)+118.74(PH_1050)+3128.51(PH_1166×1251)+2179.65(PH_1461)+92.36(PH_1606)−2294.15(PH_1251)−59.29(PH_1461×PH_1606)	2.5
%Lignin=7110.87+388.32(PH_1511×PH_1599)−16440.93(PH_1467)+447.36(PH_1599)2+19572.82(PH_1157×PH_1467)+18374.36(PH_1157)+15659.98(PH_1054×PH_1429)−4952.80(PH_1157×PH_1599)+800.20(PH_1511)−3032.75(PH_1429)2−11269.16(PH_1429)−948.04(PH_1511)2+3444.69(PH_1599)−12344.90(PH_1054)−16689.44(PH_1157)2	3.8
Note: PH – Characteristic Peak Height (Photo acoustic Units)	

PRE-TREATMENT OF LIGNOCELLULOSIC BIOMASS

Need For Pre-Treatment

Upon densification, many agricultural biomass materials, especially those from straw and stover, result in a poorly formed pellets or compacts that are more often dusty, difficult to handle and costly to manufacture. This is caused by lack of complete understanding on the natural binding characteristics of the components that make up biomass (Sokhansanj et al., 2005). The natural binding characteristics of lignocellulosic biomass can be enhanced by modifying the structure of cellulose-hemicellulose-lignin matrix by application of pre-processing and pre-treatment methods (Sokhansanj et al. 2005). It is postulated that by disrupting the lignocellulosic matrix of biomass materials via application of various chemical, physico-chemical (steam explosion, microwave, and radio frequency heating), and biological pre-treatment, the compression and compaction characteristics can be improved (Shaw 2008; Kashaninejad and Tabil, 2011). When high molecular amorphous polysaccharides are reduced to low molecular components, the polymer becomes more cohesive in the presence of moisture (Chen et al., 2004). The cellulose-hemicellulose-lignin matrix can be broken down to smaller amorphous molecules through acid or alkaline hydrolysis as well as through steam explosion (Ladisch, 1989; Vlasenko, 1997). Alkaline or acid solutions are often used for pre-treatment of biomass and the effect of pre-treatment depends on the lignin content of biomass. When biomass is treated with dilute alkaline solution, the internal surface area of the material is increased by swelling. Swelling causes a decrease in the degree of polymerization, separation of structural linkages between lignin and carbohydrates and disruption of the lignin structure (Fan et al., 1987). Increased moisture content resulting from chemical and enzymatic treatments is a problem, as the treated biomass has to be dried prior to densification. Steam explosion results in the hemicelluloses being hydrolyzed and water soluble, the cellulose is slightly depolymerized, the lignin melts and is depolymerized, which aid in binding particles together during densification. Zandersons et al. (2004) stated that activation of lignin and changes in the cellulosic structure during the steam explosion process facilitate the formation of new chemical bonds.Lam et al. (2008) reported that the quality (durability) of compacts produced from steam exploded sawdust was 20% higher than non-treated sawdust.

Physico-Chemical Pre-Treatments

Steam Explosion

Steam explosion is one of the most applied pre-treatment processes owing to its low use of chemicals and limited energy consumption (Harmsen et al., 2010). During steam explosion pre-treatment process, the lignocellulosic biomass is heated with high pressure saturated steam having temperatures typically in the range of 180-230°C for 2-10 minutes. Subsequently, the substrate is quickly flashed to atmospheric pressure; as a result, the water inside the substrate vaporizes and expands rapidly, disintegrating the biomass (Grous et al., 1985; Kokta and Ahmed, 1998; Zimbardi et al., 1999). This causes great reduction in the particle size of the substrate (Fig. 4). The heart of the explosion pulping process is the reactor, which allows the use of high pressure during heating and cooking. The reactor can be of either the batch (Fig. 5) (Jin and Chen, 2006) or continuous type (Fig. 6) (Kokta and Ahmed, 1998; Adapa et al., 2010a).

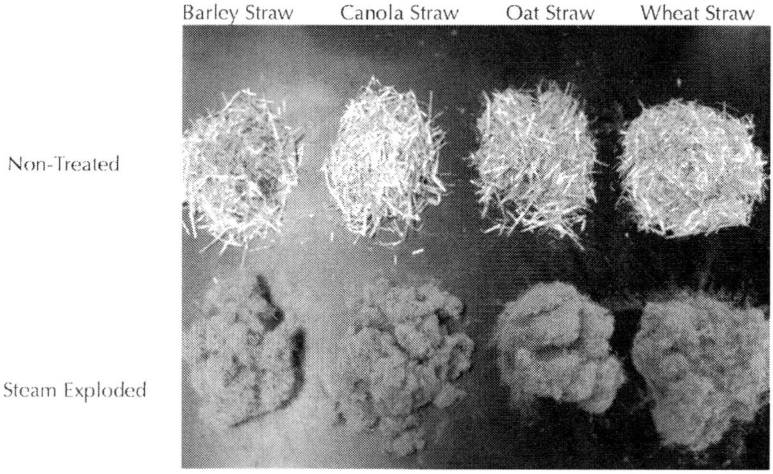

Figure 4: Photographs showing the non-treated (30 mm hammer mill screen size) and steam exploded barley, canola, oat and wheat straw grinds.

The extent of chemical and structural modifications from steam-explosion pre-treatment depends on residence time, temperature, particle size and moisture content (Sun and Cheng, 2002). However, the severity (Ro) of steam explosion is quantified as a function of retention time and reaction temperature (Equation 1) (Overend and Chornet, 1987; Viola et al. 2008).

$$R_o = t \times \exp\left(\frac{T-100}{14.75}\right) \qquad (1)$$

Where T is the temperature in °C and t is the time in minutes.

According to Zimbardi et al. (1999), the simplest way to carry out steam explosion is by batch procedure, hence widely reported in literature. However, the continuous reactors are of major interest for industrial applications. They have indicated that although the products obtained at the same treatment, severity in batch and continuous reactors are macroscopically different at first sight, there is still a lack of understanding to explain these differences. Consequently, they have developed experimental relationships between the two systems useful in making the data transfer straightforward (Equation 2).

$$\log(Ro)_{Batch} = 1.50 \times (\log(Ro)_{continuous} - 1) \quad (2)$$

In addition, studies have been carried out to try to improve the results of steam explosion by addition of chemicals such as acid or alkali (Cara et al., 2008; Harmsen et al., 2010; Stenberg et al., 1998; Zimbardi et al., 2007). Limitations of steam explosion include the formation of degradation products that may inhibit downstream processes (Garcia-Aparicio et al., 2006).

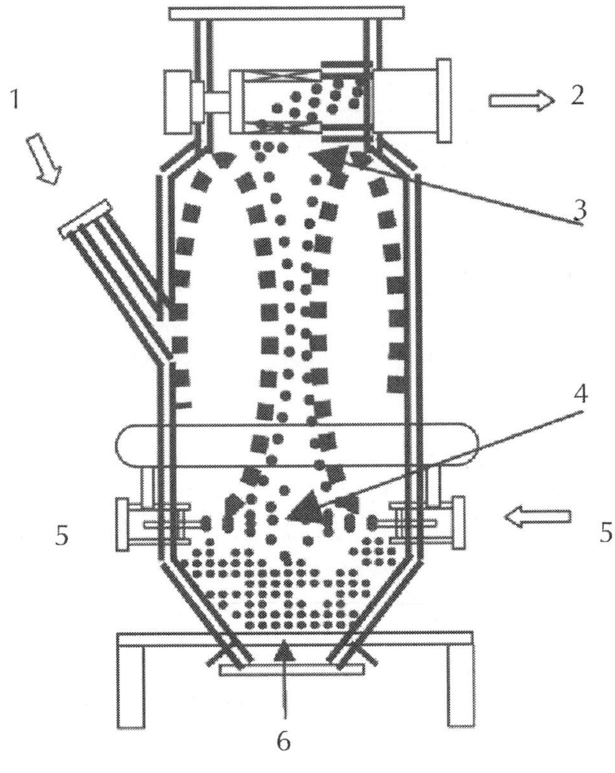

Figure 5: Schematic diagram of the FJM-200 fluidized bed opposed jet mill. 1, Infeed; 2, collection; 3, classification section; 4, grinding section; 5, compressed air; 6, discharge opening (Jin and Chen, 2006).

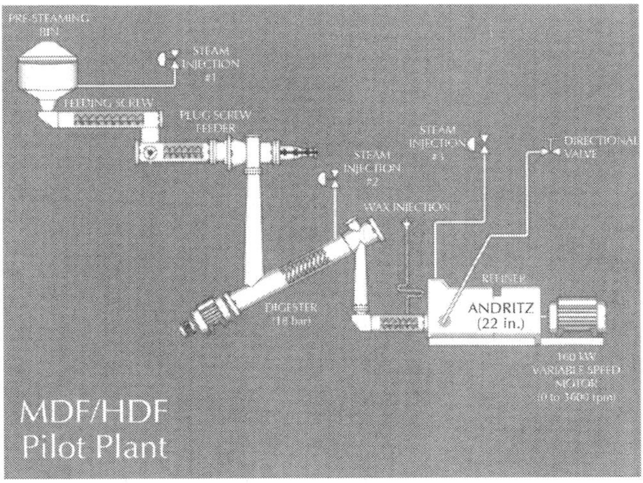

Figure 6: The And Ritz (ANDRITZ AG, Graz, Austria) continuous biomass steam explosion facility for manufacturing of Medium Density and High Density Fiberboards (MDF/HDF), Forint pilot plant at the Innovations, Quebec City, Quebec (Adapt et al., 2010a).

Microwave and Radio Frequency (RF) Heating

Dielectric heating is an alternative method to conventional heating. Unlike conduction/convection heating, which is based on superficial heat transfer, dielectric heating is based on volumetric and rapid heat transfer (de la Hoz et al., 2005). When lignocellulosic materials are placed in an electric field for dielectric heating pre-treatment, dipole molecules such as water or other dielectric materials, rotate vigorously to orient in the field. More polar components will absorb more energy, and thus, "hot spots" will be created in non-homogeneous materials. It is hypothesized that this unique heating feature results in an "explosion" effect in the particles and improves the disruption of the lignocellulosic structures. In addition, the non thermal effects of electromagnetic field accelerate the disintegration of the crystal structures (de la Hoz et al., 2005). Dielectric heating can be categorized as microwave or radio frequency depending on the wavelength used in the heating devices (Oberndorfer et al., 2000).

Microwave is electromagnetic waves between 300 MHz (wavelength 1 m) and 300 GHz (wavelength 1 mm). This range of spectrum lies between infrared and radio frequency radiation. Microwave irradiation has been extensively used in many processes because of its high heating efficiency and easy operation. Microwave energy can penetrate into materials and heat them quickly and uniformly. Microwave irradiation is considered to create thermal and non thermal effects. It has been applied as an efficient pre-treatment technique to enhance the hydrolysis of biomass materials. Some studies have demonstrated that microwave irradiation can change the structure of lignocellulosic materials and degrade or reduce lignin content, reduce cellulose crystallinity, and increase porosity and surface area of the materials (Azuma, 1984; Zhu et al., 2006b; Kashaninejad and Tabil, 2011).

This pre-treatment involves microwave irradiation of immersed biomass in an aqueous environment. Different types of lignocellulosic materials have been pre-treated using microwave irradiation including wheat straw (Ooshima et al., 1984; Zhu et al., 2006b; 2006c; Kashaninejad and Tabil, 2011), barley straw (Kashaninejad and Tabil, 2011), rice straw (Ooshima et al., 1984; Zhu et al., 2005 ; 2006a), rice hulls (Magara and Azuma, 1989), sugarcane bagasse (Ooshima et al., 1984; Magara and Azuma, 1989;Kitchaiya et al., 2003), switchgrass (Hu and Wen, 2008; Keshwani et al., 2007) and woody materials (Azuma et al., 1984). These materials were subjected to microwave pre-treatment of 2450 MHz in the range of 250 to 1000 W. The temperature of operation ranged from 70 to 230°C, while heating time varied from 5 to 120 minutes. However, higher microwave power with short pre-treatment time and the lower microwave power with long pre-treatment time had almost the same effect on chemical composition of lignocellulosic materials (Zhu et al., 2005; 2006b; Kashaninejad and Tabil, 2011).

In order to increase the efficiency of microwave heating pre-treatment, some researchers have combined microwave treatment with alkaline treatment such as NaOH or $Ca(OH)_2$. Some used alkaline solution during microwave heating treatment (Zhu et al., 2005; 2006a; 2006b; 2006c; Keshwani et al., 2007; Kashaninejad and Tabil, 2011) and some applied the alkaline solution before the lignocellulosic materials were subjected to microwave irradiation (Zhu et al., 2006a; Hu and Wen, 2008). Combination of microwave irradiation and alkali treatment improves the degradation of biomass by accelerating the reactions during the pre-treatment process compared with the conventional heating chemical pre-treatment process. Remarkable changes (Table 2) have been reported in the chemical composition of biomass samples after microwave-alkali pre-treatment, particularly in hemicellulose, lignin, and cellulose contents (Kashaninejad and Tabil, 2011). It has been reported that alkali treatment dissolves lignin and hemicellulose, and microwave irradiation facilitates dissolving these components in alkali solutions (Jackson, 1977; Kumar et al., 2009; Lesoing et al., 1980; Zhu et al., 2005). Biomass samples pretreated by microwave-alkali technique have lower lignin, hemicellulose, and cellulose than samples pretreated by microwave-distilled water or untreated samples. Moreover, degradation and depolymerisation of lignin to smaller phenolic components is another influence of microwave-alkali pre-treatment that could be considered as binder in densification process. The pellets made from microwave-chemical pre-treated biomass grinds have significantly higher density and tensile strength (Table 3) than the untreated or samples pre-treated by microwave alone (Kashaninejad and Tabil, 2011).

Radio frequency (RF) can penetrate more deeply into the materials compared with microwave heating because the radio frequency wavelength is up to 360 times greater than microwave (Marra et al., 2007). This unique characteristic is an advantage to treat large amount of material and it is easier to scale up the process. While radio frequency as a heating method has been widely applied in food-processing industries, there is not much report on application of radio frequency heating for lignocellulosic materials pre-treatment. Hu et al. (2008) used radio frequency heating in the NaOH pre-treatment of switchgrass to enhance its enzymatic digestibility. Because of the unique features of radio frequency heating (i.e., volumetric heat transfer, deep heat penetration of the samples, etc.), switchgrass could be treated on a large scale, at high solids content, and with a uniform temperature profile. At 20% solids content, radio frequency-assisted alkali pre-treatment (at 0.1 g of NaOH/g of biomass loading and 90°C) resulted in a higher xylose yield than the conventional heating pre-treatment. The optimal particle size and alkali loading in the radio frequency pre-treatment were determined to be 0.25-0.50 mm and 0.25 g of NaOH/g of biomass, respectively.

Table 2: Chemical composition (% dry basis) of untreated and microwave pretreated of wheat and barley straw at power 713 W

Treatment	Wheat straw						Barley straw					
	Protein	**Lignin**	**Ash**	**Starch**	**Cellulose**	**Hemicellulose**	**Protein**	**Lignin**	**Ash**	**Starch**	**Cellulose**	**Hemicellulose**
Untreated	1.99b	8.33a	6.33f	1.11d	44.99b	27.96a	1.61d	11.95a	6.03d	0.79c	46.93a	27.40a
Microwave-distilled water	2.24a	8.01c	8.87e	1.48b	39.69d	22.62b	2.01a	8.85b	6.28d	1.08b	45.25b	27.21a
Microwave-NaOH (1%)	1.41e	7.82d	17.32b	1.89a	35.82e	12.32d	1.80b	6.65e	16.96b	0.60e	40.81c	8.74c
Microwave-NaOH (2%)	1.36f	7.09f	34.77a	0.27f	34.77f	4.06f	1.62d	4.52f	41.43a	0.54f	35.22d	5.46d
Microwave-$Ca(OH)_2$(1%)	1.85c	8.11b	12.24d	0.69e	45.66a	14.94c	1.81b	7.27d	13.21c	0.72d	41.01c	15.00b
Microwave-$Ca(OH)_2$(2%)	1.52d	7.55e	15.89c	1.31c	42.56c	11.10e	1.68c	8.16c	16.73b	1.19a	41.24c	8.97c

Means with the same letters designation (a, b, c, d, and e) in a column are not significantly different at P = 0.05.

Table 3: Effect of microwave-chemical pre-treatments on fracture load and tensile strength of wheat and barley straw pellets at power 713 W

Treatment	**Wheat straw**		**Barley straw**	
	Fracture load (N)	**Tensile strength (MPa)**	**Fracture load (N)**	**Tensile strength (MPa)**
Untreated	19.10±5.61	0.81±0.24	16.25±5.30	0.68±0.22
Microwave-distilled water	35.00±11.93	1.48±0.46	14.25±5.31	0.61±0.21
Microwave-NaOH (1%)	85.46±22.94	3.99±0.82	57.13±12.12	2.42±0.47
Microwave-NaOH (2%)	88.00±15.86	3.69±0.66	90.75±22.42	3.59±0.98
Microwave-$Ca(OH)_2$ (1%)	67.05±19.82	3.03±0.79	42.38±10.30	1.83±0.49
Microwave-$Ca(OH)_2$ (2%)	78.25±25.07	3.31±1.03	67.25±19.94	2.88±0.91

Chemical Pre-Treatment

Different chemicals such as acids, alkalis, oxidizing agents and ozone have been used for chemical pre-treatment of lignocellulosic materials. Depending on the type of chemical used, pre-treatment could have different effects on structural components. Alkaline pre-treatment, ozonolysis, peroxide and wet oxidation pre-treatments were reportedly more effective in lignin removal, whereas dilute acid pre-treatment was more efficient in hemicellulose solubilization (Galbe and Zacchi, 2002; Sánchez and Cardona, 2008; Tomas-Pejo et al., 2008).

Acid Hydrolysis: Inorganic acids such as H_2SO_4 and HCl have been used for pre-treatment of lignocellulosic materials and have been used on a wide range of feedstocks ranging from hardwoods to grasses and agricultural residues. Acid hydrolysis can be classified as concentrated or dilute-acid hydrolysis based on the dose of acid used in the process. In the first case, the biomass is treated with high concentration of acids at ambient temperatures, which results in high conversion of lignocellulosic materials. Although concentrated acids are powerful agents for cellulose hydrolysis, they are toxic, corrosive, hazardous, and thus require reactors that are resistant to corrosion, making the pre-treatment process very expensive. In addition, the concentrated acid must be recovered after hydrolysis to make the process economically feasible (Galbe and Zacchi, 2002; Sun and Cheng, 2002).

Dilute-acid hydrolysis has been successfully developed for pre-treatment of lignocellulosic materials. Sulfuric acid at concentrations usually below 4% (wt) has been of the most interest in such studies as it is inexpensive and effective. Dilute H_2SO_4 pre-treatment can achieve high reaction rates and significantly improve cellulose hydrolysis (Esteghlalian et al., 1997). High temperature is favorable to attain acceptable rates of cellulose conversion. Despite low acid concentration and short reaction time, the use of high temperatures in dilute-acid hydrolysis accelerates the rate of hemicellulose sugar decomposition and increases equipment corrosion (Galbe and Zacchi, 2002; Taherzadeh and Karimi, 2007).

Alkali hydrolysis: Dilute alkali such as sodium, potassium, calcium, and ammonium hydroxides have been used for pre-treatment of lignocellulosic materials in alkali hydrolysis. The effectiveness of these agents depends on the lignin content of the materials. Temperature and pressure are lower in alkali pre-treatment compared with other pre-treatment methods (Mosier et al., 2005). Alkali pre-treatment can be conducted at ambient conditions, but process time is longer (hours or days instead of minutes or seconds). Compared with acid process, alkaline process causes less sugar degradation, and many of the caustic salts can be recovered and/or regenerated.

Sodium hydroxide has been studied more than other agents (Soto et al., 1994; Fox et al., 1989;MacDonald et al., 1983). Treatment of lignocellulosic materials using dilute NaOH results in swelling, leading to an increase in internal surface area, a decrease in the degree of polymerization, a decrease in crystallinity, separation of structural linkages between lignin and carbohydrates, and disruption of the lignin structure. However, calcium hydroxide (lime) is the least expensive hydroxide and has been shown to be an effective pre-treatment agent. The process of lime pre-treatment involves slurrying the lime with water, spraying it onto the biomass material, and storing the material in a pile for a period of hours to weeks. The particle size of the biomass is typically 10 mm or less. Elevated temperatures reduce contact time.

Oxidizing agents: In this pre-treatment, an oxidizing compound such as hydrogen peroxide (H_2O_2) or peracetic acid (CH_3CO_3H) is used to treat lignocellulosic materials and sometimes is applied in combination of an alkaline solution (e.g. NaOH) to improve effectiveness. This pre-treatment is usually carried out under mild temperature. This pre-treatment is more effective to increase crop residue digestibility compared with NaOH pre-treatment alone. Gould (1984) delignified agricultural residues using 1% H_2O_2 at 25°C for 18–24 h. Under this condition, more than half of the lignin and most of hemicellulose were solubilized. The pre-treatment of cane bagasse with H_2O_2 greatly enhanced its susceptibility to further hydrolysis. About 50% of the lignin and most of the hemicellulose were solubilized by 2% H_2O_2 at 30°C within 8 h, and a 95% efficiency of glucose production from cellulose was achieved in the subsequent saccharification by cellulase at 45°C for 24 h (Azzam, 1989).

Ozonolysis: In this process, ozone is used to change the structure of lignocellulosic materials and has been used for different materials such as wheat straw (Ben-Ghedalia and Miron, 1981), bagasse, green hay, peanut, pine (Neely, 1984), cotton straw (Ben-Ghedalia and Shefet, 1983) and poplar sawdust (Vidal and Molinier, 1988). Ozonolysis is carried out at room temperature and normal pressure. It can effectively remove the lignin without producing any toxic residues. In this process, hemicellulose is slightly affected, but no change in cellulose has been reported. The main restriction of this process is the large amount of ozone utilization that makes the process expensive (Sun and Cheng, 2002). Binder et al. (1980) reported 60% removal of the lignin from wheat straw using ozone pre-treatment. Enzymatic hydrolysis yield increased from 0% to 57% as the percentage of lignin decreased from 29% to 8% after ozonolysis pre-treatment of poplar sawdust (Vidal and Molinier, 1988). Garcia-Cubero et al. (2009) studied the ozonolysis pre-treatment of wheat straw in a fixed bed reactor at room conditions and concluded that enzymatic hydrolysis yield of up to 88.6% compared to 29% in non-ozonated sample.

Biological Pre-Treatment

Most pre-treatments require expensive instruments or equipment that require high energy requirements, depending on the process. In particular, physical and thermo-chemical processes require ample amount of energy to change the lignocellulosic structure of biomass. Biological pre-treatment using various types of rot fungi is a process that does not require high energy for lignin removal from a lignocellulosic biomass, despite extensive lignin degradation. Biological pre-treatments are safe, environmentally friendly and less energy intensive compared to other pre-treatment methods. However, the rate of hydrolysis reaction is very slow and needs a great improvement to be commercially applicable.

Biological pre-treatment comprises of using microorganisms such as brown-, white-, and soft-rot fungi for selective degradation of lignin and hemicellulose among which white-rot fungi seems to be the most effective microorganism (Fan et al., 1987). Brown rots mainly attack cellulose, while white and soft rots attack both cellulose and lignin. Lignin degradation occurs through the action of lignin-degrading enzymes such as peroxidases and laccase (Okano et al., 2005). These enzymes are regulated by carbon and nitrogen sources. The suitable fungi for biological pre-treatment should have higher affinity for lignin and degrade it faster than carbohydrate components.

Hatakka et al. (1983) studied the pre-treatment of wheat straw by 19 white-rot fungi and found that 35% of the straw was converted to reducing sugars by *Pleurotus ostreatus* in 5 weeks. Similar conversion was obtained in the pre-treatment by *Phanerochaete sordid* (Ballesteros et al., 2006) and*Pycnoporus cinnabarinus* (Okano et al., 2005) in 4 weeks. Akin et al. (1995) also reported the delignification of bermudagrass by white-rot fungi. The biodegradation of bermudagrass stems was improved by 29-32%, after 6 weeks, using *Ceriporiopsis subvermispora* and by 63-77% using *Cyathus stercoreus*.

PARTICLE SIZE REDUCTION AND PHYSICAL PROPERTIES

The application of pre-processing operations such as particle size reduction/ grinding is critical in order to increase the surface area of lignocellulosic biomass prior to densification (Mani et al. 2003). Particle size reduction increases the total surface area, pore size of the material and the number of contact points for inter-particle bonding in the compaction process (Drzymala, 1993). Size reduction is an important energy intensive unit operation essential for bioenergy conversion process and densification to reduce transportation costs (Bitra et al., 2009; Soucek et al., 2003). Energy consumption of grinding biomass depends on initial particle size, moisture content, material properties, feed rate of the material and machine variables (Lopo, 2002). A comprehensive summary of literature review on size reduction of lignocellulosic biomass is provided in Table 4.

Table 4: A comprehensive summary of literature review on size reduction of lignocellulosic biomass

Type of Choppers/ Grinders	**Biomass**	**Parameters Measured and Correlations**	**Observations**	**Reference**
Hammer Mill	Non-Treated and Steam Exploded Barley, Canola, Oat and Wheat Straw	Hammer mill screen size (from 1.6 to 30.0 mm) on Specific Energy Effect of geometric mean particle size on bulk density Effect of geometric mean particle size on particle density Analysis on ground particle size distribution	Negative exponential and power correlation between geometric mean particle size with both bulk and particle density Specific energy requirement is material dependent Negative power correlation between hammer mill screen size and specific energy Shapiro-Wilk Test for normality was performed	Adapa et al., 2011a
Tub Grinder	Rectangular Wheat Straw Bales, Round Rice Straw and Corn Stover Bales	Effect of screen sizes (from 12.7 to 50.8 mm) on Specific Energy Effect of screen sizes on grinding rate Analysis on ground particle size distribution	Positive correlation between screen size and grinding rate Positive correlation between tub rotational speed and grinding rate Specific Energy is material dependent Negative power/ exponential correlation between screen size and specific energy	Arthur et al., 1982

Hammer Mill	Coastal Bermudagrass	Effect of moisture content on Specific Energy Effect of feed rate on Specific Energy	Positive correlation between moisture content and specific energy Positive correlation between feed rate and specific energy	Balk, 1964
Hammer Mill	Wheat Straw and Corn Stover	Effect of operating speeds (from 2000 to 3600 rpm) on Specific Energy Effect of hammer angles (90° and 30° hammers) on Specific Energy Analysis on ground particle size distribution	Positive correlation between operating speed and specific energy Geometric mean particle diameter decreased with an increase in hammer mill speed Specific energy increased with a decrease in hammer angle	Bitra et al., 2009
Knife and Hammer Mills	Hardwood Chips, Wheat Straw and Corn Stover	Effect of screen size (from 1.6 to 12.7 mm) on specific energy	Negative correlation between screen size and specific energy Specific Energy is material dependent	Cadoche and López, 1989
Hammer Mill	Red Winter Wheat Straw	Effect of screen size (from 1.59 to 4.76 mm) on specific energy Effect of feed rate (from 1.5 to 2.5 kg/min) on specific energy	Negative correlation was observed between screen size and specific energy Feed rate did not have significant effect on specific energy	Fang et al., 1997

Hammer Mill	Wheat Straw	Effect of screen size (3.2 and 1.6 mm) on Specific Energy Analysis on ground particle size distribution	Negative correlation between specific energy and screen size	Himmel et al., 1985
Tub Grinders	Round Bales of Corn Stover and Perennial Grasses	Effect of screen size (from 19.1 to 127.0 mm) on Specific Energy Effect of screen size on throughput Effect of screen size on bulk and particle densities	Negative correlation between screen size and bulk density Positive correlation between screen size and particle density Negative correlation between screen size and specific energy Positive correlation between screen size and throughput	Kaliyan et al., 2010
Hammer Mill	Barley Straw, Corn Stover and Switchgrass	Effect of Screen Sizes (3.2, 1.6 and 0.8 mm) on Specific Energy Effect of Moisture Content (8% and 12% wb) on Specific Energy Correlation between bulk and particle densities and geometric mean diameter Analysis on ground particle size distribution	Negative linear correlation between specific energy and hammer mill screen size at 8% mc Quadratic correlation between specific energy and hammer mill screen size at 12% mc Polynomial relations for bulk and particle densities with geometric mean particle diameters	Mani et al., 2004

Hammer Mill	Oat Straw, Rattle Grass and Miscanthus	Effect of Screen size (from 1.0 to 10.0 mm) on specific energy Effect of moisture content on specific energy	Negative power correlation between screen size and specific energy Positive correlation between moisture content and specific energy; significantly higher specific energy is required at smaller screen sizes	Soucek et al., 2003
Hammer Mill	Corn Stover	Effect of hammer thickness (6.4 and 3.2 mm) on Specific Energy Effect of hammer thickness (6.4 and 3.2 mm) on Grinding Rate Effect of hammer tip speed (54 to 86 m/s) on Specific Energy	Negative correlation between hammer thickness and Specific Energy Negative correlation between hammer thickness and Grinding Rate Positive correlation between hammer tip speed and Specific Energy	Vigneault et al., 1992
Knife Mill	Switchgrass, Corn Stover, and Wheat Straw	Effect of rotational speed (from 250 to 500 rpm) on specific energy Analysis on ground particle size distribution	Positive correlation between rotational speed and specific energy Screen size has significant effect on particle size distribution	Womac et al., 2007

Chopping

Baled agricultural biomass from the field does not have good flowing characteristics and may not flow easily into grinders such as hammer mills and disc refiners. Therefore, biomass needs to be chopped with a chopper (rotary shear shredder)/ knife mill/ tub grinder to accommodate bulk flow and uniformity of feed rate. A chopper, knife cutter, or knife mill is often used for coarse size reduction (>50 mm) of stalk, straw, and grass feed stocks (Bitra et al., 2009). Knife mills reportedly worked successfully for shredding forages under various crop and machine conditions (Cadoche and López, 1989).

Bitra et al. (2009) reported that the total specific energy (including energy to operate the knife mill) for agricultural biomass chopping increases with knife mill speed. The total specific energy for knife mill and tub grinder has been observed to have negative correlation with screen size and mass feed rate (Arthur et al., 1982; Bitra et al., 2009; Himmel et al., 1985). However, grinding rate (throughput) increases with an increase in screen size (Arthur et al., 1982).

For tub grinders, an increase in screen size results in an increase in geometric mean length of particles and throughput, but a decrease in bulk density of the particles and specific energy consumption (Kaliyan et al., 2010).

Hammer Mill Grinding

Typically, hammer mills are used in forage processing industry as they are relatively inexpensive, easy to operate and produces wide range of particles (Lopo, 2002). Hammer mills have achieved merit because of their ability to finely grind a greater variety of materials than any other machines (Scholten et al., 1985). The performance of a hammer mill is measured in terms of energy consumption and geometric mean diameter and particle size distribution of the ground product (Adapa et al., 2011a; Mani et al., 2004).

Screen Size: Hammer mill screen opening size was the most significant factor affecting mill performance (Fang et al., 1997) and also has significant effect on mean particle size (Pfost and Headley, 1971). The specific energy required to grind agricultural biomass significantly increases with a decrease in hammer mill screen size and shows a negative power correlation (Arthur et al., 1982;Soucek et al., 2003). Similarly, Adapa et al. (2011a) reported negative correlation between specific energy and particle size of biomass as affected by hammer mill screen sizes. However, two other studies reported a second-order polynomial relationship between the specific energy requirements for grinding biomass (Mani et al. 2004; Sitkei, 1986). Usually, the mean geometric particle size for any particular biomass decreases with a decrease in hammer mill screen size (Adapa et al., 2011a). It has been reported that wider particle size distribution is suitable for compaction (pelleting/ briquetting) process (Adapa et al., 2011a; Mani et al., 2004). During compaction, smaller (fine) particles rearrange and fill in the void space of larger (coarse) particles producing denser and durable compacts (Tabil, 1996).

Operating Speed (Peripheral Velocity): The speed has a significant effect on mean particle size (Pfost and Headley, 1971). The total specific energy of hammer mill grinding has direct correlation to an increase in hammer tip speed (Bitra et al., 2009; Vigneault et al., 1992). High speed hammer mills with smaller diameter rotors are good for fine or hard-to-grind material. However, at high tip speeds, the material moves around the mill parallel to the screen surface making the openings only partially effective. At slower speeds, the material impinges on the screen at a greater angle causing greater amounts of coarser feed to pass through (Balk, 1964).

Hammer Angles and Thickness: The direct energy input for grinding also depends on hammer angles. In general, the specific energy for grinding decreases with an increase in hammer degrees (Bitra et al., 2009). In addition, the specific energy for grinding increases with an increase in hammer thickness (Vigneault et al., 1992).

Material Moisture Content and Feed Rate: A positive correlation has been reported between moisture content and specific energy consumption for grinding of agricultural biomass (Balk, 1964; Mani et al., 2004; Soucek et al., 2003). Feeding rate also has significant effect on specific energy consumption during hammer mill grinding and has positive correlation (O'Dogherty, 1982).

Bulk and Particle Densities, and Geometric Mean Particle Size: Usually, the bulk and particle density of agricultural straw significantly increases with a decrease in hammer mill screen size (Adapa et al., 2011a). The geometric mean particle size of pre-treated straw is usually smaller than that of the non-treated straw. This could be due to the fact that application of pre-treatment disrupts/ disintegrates the lignocellulosic structure of the biomass (Sokhansanj et al., 2005) leading to lower shear strength (easier to grind the straw).

Physical and Frictional Properties of Biomass

Bulk Density: The goal of densification is to increase the bulk density of agricultural straw to facilitate economic storage, transportation and handling of the material. In addition, densification results in an increase in the net calorific content per unit volume. The bulk density of agricultural biomass depends on the type of biomass, moisture content, grind size, and pre-treatment (Mani et al., 2006). Lower bulk densities, and concerns with uneven and low flowability of straw grinds are critical issues to sustainable production of pellets using pellet mills (Adapa et al., 2009; Larsson et al., 2008).

Typically, the bulk density of ground straw increases with a decrease in hammer mill screen size. Also, pre-treatments usually results in a decrease in bulk density since the organized lignocellulosic structure of biomass is disturbed/ disintegrated. In addition, the bulk density and geometric mean particle size of material is correlated by either power or exponential relations (Adapa et al., 2010b; Mani et al., 2004).Table 5 shows a summary of average bulk density of various agricultural biomasses ground using a hammer mill.

Particle Density: Particle size of the grinds will have direct effect on the final pellet density. Theoretically, the density of pellet can be as high as the particle density of the ground biomass. Similar to bulk density, particle density also depends on the type of biomass, moisture content, grind size, and pre-treatment (Adapa et al., 2010b). The particle density is observed to have negative correlation with hammer mill screen size. Application of pre-treatment increases the particle density since disturbance/ disintegration of lignocellulosic structure results in finer components (Adapa et al., 2010b). Table 5shows a summary of average particle density of various agricultural biomasses ground using a hammer mill.

Geometric Mean Particle Size and Distribution: It has been reported that wider particle size distribution is suitable for compaction (pelleting/ briquetting) process (Mani et al., 2004a). During compaction, smaller (fine) particles rearrange and fill in the void space of larger (coarse) particles producing denser and durable compacts (Tabil, 1996). Therefore, ideally the grinds should be normally distributed, should have near zero skewness and lower peak than expected for the normal and wider distribution of data (negative Kurtosis values). In addition, a decrease in the biomass grind size has been observed to have a positive effect on pellet mill throughput (Adapa et al., 2004).

Frictional Properties: Prior to densification, biomass grinds need to be efficiently stored, handled and transported. Physical and frictional properties of biomass have significant effect on design of new and modification of existing bins, hoppers and feeders (Fasina et al., 2006). The frictional behavior of biomass grinds in all engineering applications is described by two independent parameters: the coefficient of internal friction, and the coefficient of wall friction. The former determines the stress distribution within particles undergoing strain, and the latter describes the magnitude of the stresses between the particle and the walls of its container (Seville et al., 1997). The classic law of friction states that frictional force is directly proportional to the total force that acts normal to the shear surfaces (Chancellor, 1994; Chung and Verma, 1989; Larsson, 2010). Frictional force depends on the nature of the materials in contact but is independent of the area of contact or sliding velocity (Mohsenin, 1970). Material properties such as moisture content and particle size affect the frictional properties and densification performance of an individual feedstock (Larsson 2010; Shaw and Tabil, 2006). In addition, the determination of coefficient of friction is essential for the design of production and handling equipment and in storage structures (Adapa et al., 2010a; Puchalski and Brusewitz, 1996). A comprehensive summary of literature review on coefficient of internal friction and cohesion of agricultural biomass is provided in Table 6.

Predominantly, a linear correlation exists between normal and shear stress for agricultural straw grinds (Adapa et al., 2010a; Chevanan et al., 2008; Richter, 1954) at any specific hammer mill screen size. An increase in hammer mill screen size significantly decreases the shear stress for ground straw at any specific normal stress (Adapa et al., 2010a).

The correlation for coefficient of internal friction and cohesion with average geometric mean particle sizes for agricultural straw grinds is provided in Adapa et al. (2010a). These correlations can be used to predict the coefficient of internal friction (slope of the linear plot) and the cohesion (intercept of the linear plot) for various geometric mean particle sizes. In general, the coefficient of internal friction for ground agricultural straw decreases with an increase in average geometric mean particle diameter. The coefficient of cohesion for straw grinds increases with an increase in average geometric mean particle size (Adapa et al., 2010a).

Table 5: Average bulk and particle densities of various agricultural biomasses ground using a hammer mill

Biomass	**Hammer Mill Screen Size (mm)**	**Moisture Content (%, wb)**	**Geometric Mean Particle Diameter (mm)**	**Bulk Density (kg/m³)**	**Particle Density (kg/m³)**	**Reference**
Barley	6.4	8.9	0.88	96	1046	Adapa et al., 2011a
	3.2	5.3	0.46	149	1089	
	1.6	7.8	0.46	155	1149	
Canola	6.4	12.6	0.89	144	1019	Adapa et al., 2011a
	3.2	9.2	0.52	190	1192	
	1.6	8.3	0.37	203	1309	
Corn Stover	3.2	6.22	0.41	131	1170	Mani et al., 2004
	1.6	6.22	0.26	156	1330	
	0.8	6.22	0.19	158	1340	
Oat	6.4	10.9	0.94	111	873	Adapa et al., 2011a
	3.2	9.4	0.57	156	1093	
	1.6	7.7	0.40	196	1240	
Wheat	6.4	9.5	0.99	107	1078	Adapa et al., 2011a
	3.2	9.5	0.72	141	1225	
	1.6	8.6	0.45	154	1269	

Table 6: A comprehensive summary of literature review on coefficient of internal friction and cohesion of agricultural biomass

Biomass	**Experimental Conditions**	**Observations**	**Reference**
Corn Stover	Normal stress from 10 to 200 kPa Hammer mill screen sizes of 6.35 and 3.18 mm Moisture contents of 7, 11 and 15% (wb) Galvanized steel	Coefficient of wall friction increased from 0.18 to 0.26 with an increase in moisture from 7 to 15% No clear trend observed for the adhesion coefficient	Mani et al., 2004
Peat hull, Switchgrass and Poultry Litter	Consolidating stress from 1.5 to 12.0 kPa Screen sizes from 0.79 to 3.20 mm Ring shear test	No effect of screen size on internal friction and cohesive properties	Fasina et al., 2006
Peat Moss, Wheat Straw, Oat Hulls and Flax Shives	Normal stress from 10 to 400 kPa Geometric mean particles sizes of 0.74 (peat moss), 0.65 (wheat straw), 0.47 (oat hulls) and 0.64 (flax shives) mm Moisture content 9-10% (wb) Mild steel surface	Coefficient of wall friction were 0.68 (peat moss), 0.45 (wheat straw), 0.39 (oat hulls), and 0.41 (flax shives) Adhesion coefficients were 0.2635 kPa (peat moss), 10.687 kPa (wheat straw), 4.719 kPa (oat hulls), and 16.203 kPa (flax shives)	Shaw and Tabil, 2006

Alfalfa, Barley Straw, Wheat Straw and Whole Green Barley	Normal stress from 200 to 735 kPa Moisture content for alfalfa, barley straw from 12.0 to 45.7%, and for wheat straw at 10.0% and whole green barley at 51.0% (wb) Chop size from 10 to 90 mm Polished steel surface	Coefficient if friction on steel surface for alfalfa and barley straw increased with moisture content and was from 0.15 to 0.26, and 0.14 and 0.27, respectively Coefficient of friction for wheat straw and whole green barley were 0.13 and 0.21, respectively No effect of chop size on coefficient of internal friction on barley straw	Afzalinia and Roberge, 2007
Switchgrass, Wheat straw, and Corn Stover	Normal stresses from 1.23 to 4.92 kPa Chop size of 7.81 and 13.50 mm for switchgrass, 7.09 and 10.39 mm for wheat straw, 7.80 and 14.89 mm for corn stover	No effect of chop size on friction coefficients Coefficient of internal friction varied from 0.765 to 1.586	Chevanan et al., 2008

Reed Canary Grass	Normal stresses of 0.52-7.52 kPa (low) and 23-275 MPa (high) Moisture contents from 6.7 to 17.1% (wb) for low normal stress, and 8.9 to 27.2% for high normal stress Screen sizes of 4.0 mm Ring shear test	High friction value of 0.6 was obtained at normal stress of 50 MPa and lower At low normal stresses, the coefficient of kinematic wall friction (ratio of shear stress and normal stress) was positively correlated with moisture content and negatively correlated to normal stress At high normal stresses, the coefficient of kinematic wall friction was negatively correlated to both moisture content and normal stress	Larsson, 2010

Non-Treated and Steam Exploded Barley, Canola, Oat and Wheat Straw	Normal stress from 9.8 to 39.2 kPa Moisture content of 10% (wb) Screen size from 1.6 to 6.4 mm	No effect of screen size on shear stress values Steam exploded straw had higher coefficient of internal friction than non-treated straw grinds Coefficient of friction for non-treated barley, canola, oat and wheat straw were in the range of 0.505 to 0.584, 0.661 to 0.665, 0.498 to 0.590, and 0.532 to 0.591, respectively. Coefficient of friction for steam exploded barley, canola, oat and wheat straw were in the range of 0.562 to 0.738, 0.708 to 0.841, 0.660 to 0.860, and 0.616 to 1.036, respectively	Adapa et al., 2010a

SUMMARY

The current chapter has explored the effects of pre-treatment (chemical, physico-chemical, and biological) and pre-processing (size reduction) techniques on densification of agricultural straw resulting in high quality (density and durability) pellets. It has been determined that an increase in bulk density of biomass also increases the net calorific content per unit volume of pellets, and facilitates easy and economical storage, transport and handling of the biomass. Pre-treatment and pre-processing methods disintegrate the basic lignocellulosic structure of biomass, and change the relative composition of lignin, cellulose and hemicelluloses in the material. In addition, physical and frictional properties of agricultural straw are altered. The application of pre-treatments breaks the long-chain hydrogen bond in cellulose, making hemicelluloses amorphous, and loosening the lignin out of the lignocellulosic matrix, resulting in better quality (physically) pellets. During this process, the high molecular amorphous polysaccharides are reduced to low molecular components to become more cohesive in the presence of moisture during densification process. Particle size reduction increases the total surface area, pore size of the material and the number of contact points for inter-particle bonding in the compaction process.

It has been shown that the Fourier Transform Infrared Spectroscopy (FTIR) can be used to rapidly characterize and quantify cellulose-hemicellulose-lignin composition prior to and after application of various pre-processing and pre-treatment methods. Regression equations were developed to predict the lignocellulosic content of agricultural biomass using pure cellulose, hemicelluloses and lignin as reference samples.

REFRENCES

1. Adapa, P.K., Tabil L.G. & Schoenau, G.J.2011aGrinding Performance and Physical Properties of Non-Treated and Steam Exploded Barley, Canola, Oat and Wheat Straw. Journal of Biomass and Bioenergy,doi:10.1016/j.biombioe.2010.10.00435,549 561

2. P. K. Adapa, L. G. Tabil, G. J. Schoenau, T. Canam, T. Dumonceaux, 2011bQuantitative Analysis of Lignocellulosic Components of Non-Treated and Steam Exploded Barley, Canola, Oat and Wheat Straw using Fourier Transform Infrared Spectroscopy. The Journal of Agricultural Science and Technology, 5(7), (In-Press).
3. Adapa, P.K., Tabil L.G. & Schoenau, G.J.; 2010aPhysical and Frictional Properties of Ground Non-Treated and Steam Exploded Barley, Canola, Oat and Wheat Straw Grinds. Journal of Powder Technology,doi:10.1016/j.powtec.2010.03.038,201,230 241
4. P. K. Adapa, L. G. Tabil, G. J. Schoenau, 2010bPelleting Characteristics of Selected Biomass with and without Steam Explosion Pre-treatment. International Journal of Agricultural and Biological Engineering, 3(3), 62 79
5. P. K. Adapa, C. Karunakaran, L. G. Tabil, G. J. Schoenau, 2009Potential Applications of Infrared and Raman Spectromicroscopy for Agricultural Biomass. Agricultural Engineering International: the CIGR Ejournal, Manuscript 1081, XI(February), 1 25
6. P. K. Adapa, G. J. Schoenau, L. G. Tabil, E. A. Arinze, A. Singh, A. K. Dalai, 2007Customized and value-added high quality alfalfa products- a new concept. Agricultural Engineering International: the CIGR Ejournal, Manuscript FP 07 003, IX(June), 1 28
7. P. K. Adapa, L. G. Tabil, G. J. Schoenau, S. Sokhansanj, 2004 Pelleting characteristics of fractionated sun-cured and dehydrated alfalfa grinds Applied Engineering in Agriculture ASAE Transactions, 20(6), 813 820
8. S. Afzalinia, M. Roberge, 2007 Physical and mechanical properties of selected forage materials Canadian Biosystems Engineering 2
9. D. E. Akin, L. L. Rigsby, A. Sethuraman, W. H. Morrison, G. R. Gamble, K. E. L. Eriksson, 1995 Alterations in structure, chemistry, and biodegradability of grass lignocellulose treated with the white rot fungi Ceriporiopsis subvermispora and Cyathus stercoreus. Applied Environmental Microbiology, 61, 1591 1598
10. M. N. Anglès, F. Ferrando, X. Farriol, J. Salvadó, 2001 Suitability of steam exploded residual softwood for the production of binderless panels. Effect of the pre-treatment severity and lignin addition Biomass and Bioenergy, 21, 211 224

11. J. F. Arthur, R. A. Kepner, J. B. Dobie, G. E. Miller, P. S. Parsons, 1982 Tub grinder performance with crop and forest residues Transactions of the ASAE 1488 1494

12. J. I. Azuma, F. Tanaka, T. Koshijima, 1984Enhancement of enzymatic susceptibility of lignocellulosic wastes by microwave radiation. Journal of Fermentation Technology, 62(4), 377 384

13. M. Azzam, 1989Pre-treatment of cane bagasse with alkaline hydrogen peroxide for enzymatic hydrolysis of cellulose and ethanol fermentation. Journal of Environmental Science and Health, B., 24 (4), 421 433

14. M. O. Bagby, 1982In International Symposium on Ethanol from Biomass. H. E. Duckworth and E. A. Thompson, eds. 561 569Winnipeg, MB: The Royal Society of Canada.

15. J. J. Balatinecz, 1983The Potential of Densification in Biomass Utilization. In: W.A. Cote. ed. Biomass utilization. London: Plenum Press: 181 189p.

16. W. A. Balk, 1964Energy requirements for dehydrating and pelleting coastal Bermuda grass. Transactions of the ASAE, 4, 349 341

17. D. Ben-Ghedalia, J. Miron, 1981 The effect of combined chemical and enzyme treatment on the saccharification and in vitro digestion rate of wheat straw Biotechnology and Bioengineering 823 831

18. D. Ben-Ghedalia, G. Shefet, 1983Chemical treatments for increasing the digestibility of cotton straw. Journal of Agricultural Sciences, 100, 393 400

19. S. Bhattacharya, S. Sett, R. M. Shrestha, 1989 State of the Art for Biomass Densification Energy Sources 161 182

20. A. Binder, L. Pelloni, A. Fiechter, 1980 Delignification of straw with ozone to enhance biodegradability Applied Microbiology and Biotechnology, 11, 1 5

21. Bitra, V.S.P., Womac, A.R., Chevanan, N., Miu, P.I., Igathinathane, C., Sokhansanj, S., & Smith, D.R.2009Direct mechanical energy measured of hammer mill comminution of switchgarss, wheat straw, and corn stover and analysis of their particle size distributions. Powder Technology, 193(2009)32 45

22. B. O. Budevska, 2002Handbook of Vibrational Spectroscopy, 5 edited by J. M. Chalmers and P. R. Griffiths, New York: John Wiley and Sons: 3720-3732 p.
23. L. Cadoche, G. D. López, 1989 Assessment of size reduction as a preliminary step in the production of ethanol for lignocellulosic wastes. Biological Wastes 153 157
24. C. Cara, E. Ruiz, M. Ballesteros, P. Manzanares, Negro. M. Jose, E. Castro, 2008 Production of fuel ethanol from steam-explosion pretreated olive tree pruning Fuel 692 700
25. Chabannes, M., Ruel, K., Yoshinaga, A,. Chabbert, B., Jauneau, A., Joseleau, J.P., & Boudet, A.M.2001In situ analysis of lignins in transgenic tobacco reveals a differential impact of individual transformations on the spatial patterns of lignin deposition at the cellular and subcellular levels. Plant Journal, 28,doi:10.1046/j.1365-313X.2001.01159.x 271 282
26. W. L. Chancellor, 1994Friction between soil and equipment materials. ASAE Paper 941034St. Joseph, MI.: American Society of Agricultural and Biological Engineers.
27. W. Chen, G. C. Lickfield, C. Q. Yang, 2004 Molecular modeling of cellulose in amorphous state. Part I: model building and plastic deformation study Polymer 1063 1071
28. N. Chevanan, A. R. Womac, V. S. Bitra, D. C. Yoder, 2008 Flowability parameters for chopped switchgrass, wheat straw and corn stoverASABE Annual International Meeting, Rhode Island, Paper 084162St. Joseph, MI: American Society of Agricultural and Biological Engineers, June 29 July 2, 12 pages.
29. L. Chico-Santamarta, A. Humphroes, D. White, K. Chaney, R. J. Godwin, 2009Effect of Pre- and Post-Pelletization Storage of Canola (Oilseed Rape) Straw on the Quality and Properties of Pellets. ASABE Annaul International Meeting, Grand Sierra Resort and Casino, Reno, Nevada, Paper 096105June 21 24 pages.
30. J. H. Chung, L. R. Verma, 1989Determination of friction coefficients of beans and peanuts. Transactions of the ASAE 32(2), 745
31. R.L.Crawford,1981 LigninbiodegradationandtransformationNew York: John Wiley and Sons. 0-47105-743-6

32. A. de la Hoz, A. Diaz-Ortiz, A. Moreno, 2005 Microwaves in organic synthesis: Thermal and non-thermal microwave effects Chemical Society Reviews 164 178

33. S. Deguchi, K. Tsujii, K. Horikoshi, 2006 Cooking Cellulose in Hot and Compressed Water Chemical Communications 3293 3295 doi:10.1039/b605812d

34. A. Demirba , 2001Biomass Resource Facilities and Biomass Conversion Processing for Fuels and Chemicals. Energy Conversion and Management, 41, 1357 1378

35. Z. Drzymala, 1993Industrial briquetting- fundamentals and methods. Studies in mechanical engineering, 13Warszawa: PWN-Polish Scientific Publishers.

36. Encyclopædia Britannica. (2008). Cellulose. Retrieved May 23, 2008, from Encyclopædia Britannica Online.

37. A. Esteghlalian, A. G. Hashimoto, J. J. Fenske, M. H. Penner, 1997Modeling and optimization of the dilute-sulfuric-acid pre-treatment of corn stover, poplar and switchgrass. Bioresource Technology, 59, 129 136

38. L. T. Fan, M. M. Gharpuray, Y. H. Lee, 1987In Cellulose Hydrolysis Biotechnology Monographs, 57, 149 187New York, NY: Springer-Verlag.

39. Q. Fang, I. Bolloni, E. Haque, C. K. Spillman, 1997 Comparison of energy efficiency between a roller mill and a hammer mill Applied Engineering in Agriculture 631 635

40. O. O. Fasina, D. Bransby, C. W. Wood, 2006Flow properties of biomass feedstock. ASABE Annual International Meeting, Portland, Oregon, Paper 066149July 9 12pages.

41. B. Focher, A. Marzetti, E. Marsano, G. Conio, A. Tealdi, A. Cosani, M. Terbojevich, 1998Regenerated and Graft Copolymer Fibres from Steam-Exploded Wheat Straw: Characterization and Properties. Journal of Applied Polymer Science, 67, 961 974

42. D. J. Fox, P. P. Gray, N. W. Dunn, L. M. Warwick, 1989Comparison of alkali and steam (acid) pre-treatments of lignocellulosic materials to increase enzymic susceptibility: Evaluation under optimized pre-treatment conditions. Journal of Chemical Technology & Biotechnology, 44, 135 146

43. M. Galbe, G. Zacchi, 2002 A review of the production of ethanol from softwood. Applied Microbiology and Biotechnology 618 628

44. M. T. Garcia-Cubero, G. Gonzalez-Benito, I. Indacoechea, M. Coca, S. Bolado, 2009Effect of ozonolysis pre-treatment on enzymatic digestibility of wheat and rye straw. Bioresource Technology, 100, 1608 1613

45. M. P. Garcia-Aparicio, I. Ballesteros, A. Gonzalez, J. M. Oliva, M. Ballesteros, M. J. Negro, 2006Effect of inhibitors released during steam-explosion pre-treatment of barley straw on enzymatic hydrolysis. Applied Biochemistry and Biotechnology, 129(1-3), 278 288

46. Gelbrich, J., Mai, C. & Militz, H.2009Evaluation of Bacterial Wood Degradation by Fourier-Transform-Infrared (FTIR) Measurements, The study has been done within the EU-project BACPOLES (EVK4-CT-2001-00043) at the University of Gottingen (2009), 6 pages4

47. J. M. Gould, 1984 Alkaline peroxide delignification of agricultural residues to enhance enzymatic saccharification Biotechnology and Bioengineering 46 52

48. E. Granada, González. L. M. López, J. L. Míguez, J. Moran, 2002 Fuel lignocellulosic briquettes, die design and products study Renewable Energy 561 573

49. Grous, W.R., Converse, A.O., & Grethlein, H.E.1985Effect of Stem Explosion Pre-treatment on Pore Size and Enzymatic Hydrolysis or Poplar. Enzyme Microbial Technology, 1986(8), 274 280

50. Harmsen, P.F.H., Huijgen, W.J.J., Bermudez Lopez, L.M., & Bakker, R.R.C.2010Literature Review of Physical and Chemical Pre-treatment Processes for Lignocellulosic Biomass. Energy Research Centre of the Netherlands. Report accessed on January 2010. Web Address:http://www.ecn.nl/docs/library/report/2010/e10013.pdf

51. A. I. Hatakka, 1983Pre-treatment of wheat straw by white-rot fungi for enzymic saccharification of cellulose. Applied Microbiology and Biotechnology, 18, 350 357

52. M. Himmel, M. Tucker, J. Baker, C. Rivard, K. Oh, K. Grohmann, 1985Comminution of biomass: Hammer and knife mills. Biotechnology and Bioengineering Symposium, 15, 39 58

53. D. N.S. Hon, 1989Cellulosic adhesives. In Adhesives from Renewable Resources, Eds. R.W. Hemmingway and A.H. Conner, 289 304Washington, DC: American Chemical Society.

54. Z. Hu, Z. Wen, 2008Enhancing enzymatic digestibility of switchgrass by microwave assisted alkali pre-treatment. Biochemical Engineering Journal, 38, 369 378

55. Z. Hu, Y. Wang, Z. Wen, 2008Alkali (NaOH) Pre-treatment of switchgrass by radio frequency-based dielectric heating. Applied Biochemistry and Biotechnology, 148, 71 81

56. Igathinathane, C., Womac, A.R., Sokhansanj, S., & Narayan, S.2008 Bioresource Technology Knife grid size reduction to pre-process packed beds of high- and low-moisture switchgrass.,99,2254 2264

57. M. G. Jackson, 1977The alkali treatment of straws. Animal Feed Science and Technology 105 130

58. Jin, S. & Chen, H.2006 Biochemical Engineering Journal Superfine Grinding of Steam-Exploded Rice Straw and its Enzymatic Hydrolysis.,30,225 230

59. N. Kaliyan, D. R. Schmidt, R. V. Morey, 2010 Commercial Scale Grinding of Corn Stover and Perennial GrassesASABE Annual International Meeting, David L. Lawrence Convention Centre, Pittsburgh, Pennsylvania, June 20 23 Paper 1009062Niles Road, St. Joseph, MI 49085-9659 USA.

60. N. Kaliyan, R. V. Morey, 2006 Factors affecting strength and durability of densified products. ASABEAnnual International Meeting, American Society of Agricultural and Biological Engineers, Portland, Oregon July 9 12 Paper 066077Niles Road, St. Joseph, MI 49085-9659 USA.

61. M. Kashaninejad, L. G. Tabil, 2011 Effect of microwave-chemical pre-treatment on compression characteristics of biomass grinds Biosystems Engineering 36 45

62. Kelley, S.S., Rowell, R.M., Davis, M., Jurich, C.K., & Ibach, R.2004Rapid Analysis of the Chemical Composition of Agricultural Fibres Using Near Infrared Spectroscopy and Pyrolysis Molecular Beam Mass Spectrometry. Biomass and Bioenergy, 27(2004)77 88

63. D. R. Keshwani, J. J. Cheng, J. C. Burns, L. Li, V. Chiang, 2007Microwave pre-treatment of switchgrass to enhance enzymatic hydrolysis. ASABE Annual Meeting Paper 077127Minneapolis, MN, USA.
64. P. Kitchaiya, P. Intanakul, M. Krairiksh, 2003Enhancement of enzymatic hydrolysis of lignocellulosic wastes by microwave pre-treatment under atmospheric pressure. Journal of Wood Chemistry and Technology, 23(2), 217 225
65. B. V. Kokta, A. Ahmed, 1998Chapter 6: Steam Explosion Pulping. Environmentally Friendly Technologies for the Pulp and Paper Industry, edited by Raymond A. Young and Masood Akhtar, 0-47115-770-8Wiley and Sons, Inc.
66. P. Kumar, D. M. Barrett, M. J. Delwiche, P. Stroeve, 2009Methods for pre-treatment of lignocellulosic biomass for efficient hydrolysis and biofuel production. Industrial and Engineering Chemistry Research, 48, 3713 3729
67. M. R. Ladisch, 1989Hydrolysis. In: Biomass Handbook. O. Kitani and C. W. Hall, eds. 434 451New York: Gordon and Breach.
68. P. S. Lam, S. Sokhansanj, X. Bi, C. J. Lim, S. Mani, S. Melin, 2008Energetic feasibility of biomass densification with steam explosion pretreatment. CSBE Annual General Meeting; July 14 16Poster CSBE08305Vancouver, BC.
69. S. Larkin, J. Ramage, J. Scurlock, 2004Bioenergy. In: G. Boyle. ed. Renewable energy. Power for a sustainable future: 106 146p.
70. S. H. Larsson, 2010 Kinematic Wall Friction Properties of Reed Canary Grass Powder at High and Low Normal Stresses Powder Technology 108 113
71. Larsson, S.H., Thyrel, M., Geladi, P., & Lestander, T.A.2008 Bioresource Technology High Quality Biofuel Pellet Production from Pre-Compacted Low Density Raw Materials.,99,7176 7182
72. Lebo, S. E. Jr., Gargulak, J.D., & McNally, T.J.2001"Lignin". Kirk-Othmer Encyclopedia of Chemical Technology. John Wiley & Sons, Inc.DOI:10.1002/0471238961.12090714120914.a01.pub2Retrieved on 2007-10-14
73. P. Lehtikangas, 2001Quality properties of palletized sawdust, logging residues and bark. Biomass and Bioenergy, 20, 351 360

74. G. Lesoing, T. Klopfenstein, I. Rush, J. Ward, 1980 Chemical treatment of wheat straw Journal of Animal Science, 51, 263 269

75. Y. Lin, S. Tanaka, 2006 Ethanol fermentation from biomass resources: current state and prospects Applied Microbiology and Biotechnology 627 642

76. R. Liu, H. Yu, Y. Huang, 2005 Structure and Morphology of Cellulose in Wheat Straw Cellulose 25 34

77. P. Lopo, 2002The right grinding solution for you: roll, horizontal or vertical. Feed Management, 53(3), 203 206

78. Luypaert, J., Zhang, M.H., & Massart, D.L.2003Feasibility Study for the Use of Near Infrared Spectroscopy in the Qualitative and Quantitative Analysis of Green Tea. Analytca Chimica Acta, 478(2003), 303 312

79. MacDonald, D.G., Bakhshi, N.N., Mathews, J.F., Roychowdhury, A., Bajpai, P., & Moo-Young, M.1983 Biotechnology and Bioengineering Alkali treatment of corn stover to improve sugar production by enzymatic hydrolysis,25,2067 2076

80. K. Magara, J. Azuma, 1989Microwave-irradiation of lignocellulosic materials X: Conversion of microwave-irradiated agricultural wastes into ethanol. Wood Research, 76, 1 9

81. Mani, S., Tabil, L.G., & Sokhansanj, S.2006 Biomass and Bioenergy, Effects of Compressive Force, Particle Size, and Moisture Content on Mechanical Properties of Biomass Pellets from Grasses.,30,648 654

82. Mani, S., Tabil, L.G., & Sokhansanj, S.2004 Biomass and Bioenergy Grinding performance and physical properties of wheat and barley straws, corn stover and switchgrass.,27,339 352

83. S. Mani, L. G. Tabil, S. Sokhansanj, 2003An overview of compaction of biomass grinds. Powder Handling & Process, 15(3), 160 168

84. F. Marra, J. Lyng, V. Romano, B. Mc Kenna, 2007 Radio-frequency heating of foodstuff: Solution and validation of a mathematical model Journal of Food Engineering 998 1006

85. J. Mc Mullen, O. O. Fasina, C. W. Wood, Y. Feng, 2005 Storage and handling characteristics of pellets from poultry litter Applied Engineering in Agriculture 645 651

86. N. N. Mohsenin, 1970 Physical Properties of Plant and Animal MaterialsNew York, N.Y.: Gordon and Breach Science Publishers.

87. N. Mosier, C. Wyman, B. Dale, R. Elander, Y. Y. Lee, M. Holtzapple, M. Ladisch, 2005Features of promising technologies for pre-treatment of lignocellulosic biomass. Bioresource Technology, 96, 673 686

88. J. D. Murphy, K. Mc Carthy, 2005 Ethanol production from energy crops and wastes for use as a transport fuel in Ireland Applied Energy 148 166

89. W. C. Neely, 1984Factors affecting the pre-treatment of biomass with gaseous ozone. Biotechnology and Bioengineering, 26, 59 65

90. D. L. Nelson, M. M. Cox, 2005 Lehninger Principles of BiochemistryNew York, NY: W.H. Freeman and Company.

91. I. Obernberger, G. Thek, 2004Physical characterization and chemical composition of densified biomass fuels with regard to their combustion behavior. Biomass and Bioenergy, 27, 653 669

92. C. Oberndorfer, E. Pawelzik, W. Lucke, 2000 Prospects for the application of dielectric heating processes in the pre-treatment of oilseeds European Journal of Lipid Science and Technology 487 493

93. M. J. O'Dogherty, 1982 A review of research on forage chopping Journal of Agricultural Engineering Research, 27(4), 267 289

94. K. Okano, M. Kitagawa, Y. Sasaki, T. Watanabe, 2005 Conversion of Japanese red cedar (Cryptomeria japonica) into a feed for ruminants by white-rot basidiomycetes Animal Feed Science and Technology 235 243

95. H. Ooshima, K. Aso, Y. Harano, T. Yamamoto, 1984 Microwave treatment of cellulosic materials for their enzymatic hydrolysis Biotechnology Letters 289 294

96. R. P. Overend, E. Chornet, 1987Fractionation of lignocellulosics by steam-aqueous pre-treatment. Philosophical Transactions of the Royal Society, London A 321, 523 536

97. H. B. Pfost, V. E. Headley, 1971Use of logarithmic normal distribution to describe hammer mill performance. Transactions of the ASABE, 14(3), 531 535

98. C. Puchalski, G. H. Brusewitz, 1996 Coefficient of friction of watermelon Transactions of the ASABE, 39(2), 589 594

99. Ó. J. Sánchez, C. A. Cardona, 2008 Trends in biotechnological production of fuel ethanol from different feedstocks Bioresource Technology 5270 5295

100. D. J. Schell, C. Harwood, 1994 Milling of Lignocellulosic Biomass: Results of Pilot-Scale Testing Applied Biochemistry and Biotechnology 159 168

101. R. L. Scholten, R. R. Mc Ellhiney, 1985The effects of prebreaking in hammer mill particle size reduction. American Society of Agricultural Engineers; Paper 85-3542 85 3542St. Joseph MI.

102. J. P. K. Seville, U. Tuzun, R. Clift, 1997 Processing of Particulate SolidsLondon, U.K.: Chapman and Hall.

103. T. Shambe, J. F. Kennedy, 1985Acid and enzymatic hydrolysis of chaotropically pretreated millet stalk, acha and rice straws and conversion of the products to ethanol. Enzyme and Microbial Technology, 7, 115 120

104. M. Shaw, 2008 Feedstock and Process Variables Influencing Biomass DensificationM.Sc. Thesis, Department of Agricultural and Bioresource Engineering, University of Saskatchewan. Website:http://library2.usask.ca/theses/available/etd-03092008115636 /unrestricted/Thesis.MarkShaw.FinalR3.pdf

105. M. D. Shaw, L. G. Tabil, 2006 Mechanical properties of selected biomass grindsASABE Annual International Meeting, Oregon Convention Center, Portland, Oregon, Paper 066175July 9-12: 12 pages.

106. Hsu. C. P. Sherman, 1997Chapter 15: Infrared Spectroscopy, in: F. Settle, Editor, Handbook of Instrumental Techniques for Analytical Chemistry, Prentice-Hall, NJ 270.

107. G. Sitkei, 1986 Mechanics of agricultural materialsAmsterdam: Elsevier.

108. Smola, N., & Urleb, U.2000 Analytica Chimica Acta Qualitative and Quantitative Analysis of Oxytetracycline by Near-Infrared Spectroscopy.410203 210

109. S. Sokhansanj, S. Mani, X. Bi, P. Zaini, L. G. Tabil, 2005 Binderless pelletization of biomassASAE Annual International Meeting,

Tampa Convention Centre, Tampa, Florida; July 17 20Paper 056061Niles Road, St. Joseph, MI 49085-9659 USA.

110. M. L. Soto, H. Dominguez, M. J. Nunez, J. M. Lema, 1994 Enzymatic saccharification of alkali-treated sunflower hulls Bioresource Technology 53 59

111. J. Soucek, I. Hanzlikova, P. Hulta, 2003A fine disintegration of plants suitable for composite biofuels production. Research in Agricultural Engineering, 49(1), 7 11

112. K. Stenberg, C. Tengborg, M. Galbe, G. Zacchi, 1998Optimisation of steam pre-treatment of SO2-impregnated mixed softwoods for ethanol production. Journal of Chemical Technology and Biotechnology, 71(4), 299 308

113. F. Sun, H. Chen, 2007Evaluation of enzymatic hydrolysis of wheat straw pretreated by atmos-pheric glycerol autocatalysis. Journal of chemical technology and biotechnology, 82, 1039 1044

114. Y. Sun, J. Cheng, 2002 Hydrolysis of lignocellulosic materials for ethanol production: a review. Bioresource Technology 1 11

115. L. G. Tabil, 1996Pelleting and binding characteristics of alfalfa. UnpublishedPh.D. thesis, Department of Agricultural and Bioresource Engineering, Saskatoon, SK Canada: University of Saskatchewan.

116. M. J. Taherzadeh, K. Karimi, 2007 Acid-based hydrolysis processes for ethanol from lignocellulosic materials: a review. Bioresources, 2, 472 499

117. M. Temmerman, F. Rabier, P. D. Jensen, H. Hartmann, T. Bohm, 2006 Comparative Study of Durability Test Methods for Pellets and Briquettes Biomass and Bioenergy 964

118. E. Tomas-Pejo, J. M. Oliva, M. Ballesteros, 2008 Realistic approach for full-scale bioethanol production from lignocellulose: a review Journal of Scientific & Industrial Research, 67, 874 884

119. K. Tripathi, P. V. R. Iyer, T. C. Kandpal, 1988 A Techno-Economic Evaluation of Biomass Briquetting in India Biomass and Bioenergy 479 488

120. Tucker, M.P., Mitri, R.K., Eddy, F.P., Nguyen, Q.A., Gedvilas, L.M., & Webb, J.D.2000 Applied Biochemistry and Biotechnology Fourier Transform Infrared Quantification of Sugars in Pretreated Biomass Liquors.,84-86,39 50

121. United States Department of Energy (USDE).2006. Energy Efficiency and Renewable Energy- Biomass Program. Website Accessed on January 21, 2011. Web Address: http://www1.eere.energy.gov/biomass

122. D. M. Updegraff, 1969 Semimicro Determination of Cellulose in Biological Materials. Analytical Biochemistry 420 424

123. J. E. G. van Dam, Oever. M. J. A. van den, W. Teunissen, E. R. P. Keijsers, A. G. Peralta, 2004 Process for production of high density/high performance binderless boards from whole coconut huskPart 1: lignin as intrinsic thermosetting binder resin. Industrial Crops and Products 207 216

124. Vazquez, G., Gonzalez, J., Freire, S., & Antorrena, G.2002Characterization of Eucalyptus Globules and Pinus Pinaster Acetosolv Pulps Prebleached with O2 by FTIR and DRIFT Spectroscopy. Holz als Roh- und Werkstoff, 60(2002), 25 30

125. E. Y. Vlasenko, H. Ding, J. M. Labavitch, S. P. Shoemaker, 1997 Enzymatic Hydrolysis of Pretreated Rice Straw Bioresource Technology 109 119

126. P. F. Vidal, J. Molinier, 1988 Ozonolysis of lignin-improvement of in vitro digestibility of poplar sawdust. Biomass 1 17

127. C. Vigneault, T. M. Rothwell, G. Bourgeois, 1992Hammer mill grinding rate and energy requirements for thin and conventional hammers. Canadian Biosystems Engineering, 34, 203 206

128. Viola, E., Zimbardi, F., Cardinale, M., Cardinale, G., Braccio, G., & Gambacorta, E.2008 Bioresource Technology Processing Cereal Straws by Steam Explosion in a Pilot Plant to Enhance Digestibility in Ruminants,99,681 689

129. A. R. Womac, C. Igathinathane, P. Bitra, P. Miu, T. Yang, S. Sokhansanj, S. Narayan, 2007 Biomass Pre-Processing Size Reduction with Instrumented MillsASABE Annual International Meeting, Minneapolis Convention Centre, Minneapolis, June 17 20Paper 076046Niles Road, St. Joseph, MI 49085-9659 USA.

130. J. Zandersons, J. Gravitis, A. Zhurinsh, A. Kokorevics, U. Kallavus, C. K. Suzuki, 2004 Carbon materials obtained from self-binding sugar cane bagasse and deciduous wood residues plastics Biomass and Bioenergy 345 360

131. F. Zimbardi, E. Viola, F. Nanna, E. Larocca, M. Cardinale, D. Barisano, 2007 Acid impregnation and steam explosion of corn stover in batch processes Industrial Crops and Products 195 206
132. F. Zimbardi, D. Viggiano, F. Nanna, M. Demichele, D. Cuna, G. Cardinale, 1999 Steam Explosion of Straw in Batch and Continuous Systems Applied Biochemistry and Biotechnology 117 125
133. Zhu, S., Wu, Y., Yu, Z., Zhang, X., Li, H., & Gao, M.2006a Bioresource Technology, The effect of microwave irradiation on enzymatic hydrolysis of rice straw.,97,1964 1968
134. S. Zhu, Y. Wu, Z. Yu, Q. Chen, G. Wu, F. Yu, C. Wang, S. Jin, 2006b Microwave-assisted alkali pre-treatment of wheat straw and its enzymatic hydrolysis Biosystems Engineering 437 442
135. S. Zhu, Y. Wu, Z. Yu, X. Zhang, C. Wang, F. Yu, S. Jin, 2006c Production of ethanol from microwave-assisted alkali pretreated wheat straw Process Biochemistry 869 873
136. S. Zhu, Y. Wu, Z. Yu, J. Liao, Y. Zhang, 2005Pre-treatment by microwave/alkali of rice straw and its enzymic hydrolysis. Process Biochemistry, 40, 3082 3086

Chapter 3

Biomass Feedstock Pre-Processing– Part 2: Densification

Tabil Lope[1], Adapa Phani[1], and Kashaninejad Mahdi[2]

[1] Department of Chemical and Biological Engineering, University of Saskatchewan, Canada

[2] Department of Food Science & Technology, Gorgan University of Agricultural Sciences and Natural Resources Gorgan, Iran

INTRODUCTION

The Need for Densification

Agricultural biomass residues have the potential for the sustainable production of bio-fuels and to offset greenhouse gas emissions (Campbell et al., 2002; Sokhansanj et al., 2006). Straw from crop production and agricultural residues existing in the waste streams from commercial crop processing plants have little inherent value and have traditionally constituted a disposal problem. In fact, these residues represent an abundant, inexpensive and readily available source of renewable lignocellulosic biomass (Liu et al., 2005). New methodologies need to be developed to process the biomass making it suitable feedstock for bio-fuel production. In addition, some of the barriers in the economic use of agricultural crop residue are the variable quality of the residue, the cost of collection, and problems in transportation and storage (Bowyer and Stockmann, 2001; Sokhansanj et al., 2006).

In order to reduce industry's operational cost as well as to meet the requirement of raw material for biofuel production, biomass must be processed and handled in an efficient manner. Due to its high moisture content, irregular shape and size, and low bulk density, biomass is very difficult to handle, transport, store, and utilize in its original form (Sokhansanj et al., 2005). Densification of biomass into durable compacts is an effective solution to these problems and it can reduce material waste. Densification can increase the bulk density of biomass from an initial bulk density of 40-200 kg/m^3 to a final compact density of 600-1200 kg/m^3 (Adapa et al., 2007; Holley, 1983; Mani et al., 2003; McMullen et al., 2005; Obernberger and Thek, 2004). Biomass can be compressed and stabilized to 7–10 times densities of the standard bales by the application of pressures between 400–800 MPa during the densification process (Demirbas and Sahin, 1998). Because of their uniform shape and size, densified products may be easily handled using standard handling and storage equipment, and they can be easily adopted in direct-combustion or co-firing with coal, gasification, pyrolysis, and utilized in other biomass-based conversions (Kaliyan and Morey, 2006a) such as biochemical processes. Upon densification, many agricultural biomass materials, especially those from straw and stover, result in a poorly formed pellets or compacts that are more often dusty, difficult to handle and costly to manufacture. This is caused by lack of complete understanding on the natural binding characteristics of the components that make up biomass (Sokhansanj et al., 2005).

Fuel Pellet Quality Parameters

The quality of fuel pellet is usually assessed based on its density and durability. High density of pellet represents higher energy per unit volume of material, while durability is the resistance of pellets to withstand various shear and impact forces applied during handling and transportation. High bulk density increases storage and transport capacity of pellets. Since feeding of boilers and gasifiers generally is volume-dependent, variations in bulk density should be avoided (Larsson et al., 2008). A bulk density of 650 kg/m^3 is stated as design value for wood pellet producers (Obernberger and Thek, 2004). Low durability of pellets results in problems like disturbance within pellet feeding systems, dust emissions, and an increased risk of fire and explosions during pellet handling and storage (Temmerman et al., 2006). Other quality factors of biomass for thermo-chemical conversion include (FAO, 2011; Rajvanshi, 1986):

- *Energy content:* The choice of a biomass for energy conversion will in part be decided by its heating value. The method of measurement of the biomass energy content will influence the estimate of efficiency of a given gasifier. The only realistic way of presenting fuel heating values for gasification purposes is to give lower heating values (excluding the heat of condensation of the water produced) on an ash inclusive basis and with specific reference to the actual moisture content of the fuel.
- *Moisture content:* High moisture contents reduce the thermal efficiency since heat is used to drive off the water and consequently this energy is not available for the reduction reactions and for converting thermal energy into chemical bound energy in the gas. Therefore, high moisture contents result in low gas heating values during thermo-chemical processes.
- *Volatile matter:* The amount of volatiles in the feedstock determines the necessity of special measures (either in design of the gasifier or in the layout of the gas cleanup train) in order to remove tars from the product gas in engine applications.
- *Ash content and slagging characteristics:* The mineral content in the biomass that remains in oxidation form after complete combustion is usually called ash. The ash content of a fuel and the ash composition have a major impact on trouble free

operation of a gasifier or a burner. Slagging or clinker formation in the reactor, caused by melting and agglomeration of ashes, at the best will greatly add to the amount of labour required to operate the gasifier. If no special measures are taken, slagging can lead to excessive tar formation and/or complete blocking of the reactor.

- *Reactivity:* The reactivity is an important factor determining the rate of reduction of carbon dioxide to carbon monoxide in a gasifier. Reactivity depends in the first instance on the type of fuel. For example, it has been observed that fuels such as wood, charcoal and peat are far more reactive than coal.
- *Size and size distribution:* Low bulk density feedstock may cause flow problems in the gasifier or burner as well as an inadmissible pressure drop over the reduction zone and a high proportion of dust in the gas. Large pressure drops will lead to reduction of the gas load, resulting in low temperatures and tar production. Excessively large sizes of particles or pieces give rise to reduction in reactivity of the fuel, resulting in start-up problems and poor gas quality, and to transport problems through the equipment. A large range in size distribution of the feedstock will generally aggravate the above phenomena. Too large particle sizes can cause gas channelling problems. Fluidized bed gasifiers are normally able to handle fuels with particle diameters varying between 0.1 and 20 mm (FAO, 2011).
- *Bulk density:* Fuels with high bulk density are advantageous because they represent a high energy-for-volume value. Consequently, these fuels need less bunker space for a given refuelling time. Low bulk density fuels sometimes give rise to insufficient flow under gravity, resulting in low gas heating values and ultimately in burning of the char in the reduction zone. Inadequate bulk densities can be improved by briquetting or pelletizing.

All of the abovementioned biomass properties could be altered by subjecting raw biomass to various processing methods and forming composites. Before choosing a gasifier, it is important to ensure that the individual biomass meets the requirements of the gasifier or that it can be treated to meet these requirements.

Effect of Moisture Content on Pellet Quality

The moisture in biomass both acts as a facilitator of natural binding agents and a lubricant (Kaliyan and Morey, 2006a). Many studies have indicated that the production of high quality pellets is possible only if the moisture content of the feed is between 8 and 12% (wb). Moisture contents above or below this range would lead to lower quality pellets (Hill and Pulkinen, 1988; Kashaninejad et al., 2011; Li and Liu, 2000; Obernberger and Thek, 2004; Shaw and Tabil, 2007). In general, an increase in moisture content from 10 to 44% could result in up to 30-40% decrease in pellet densities of biomass (Chancellor, 1962; Grover and Mishra, 1996; Gustafson and Kjelgaard, 1963; Kaliyan and Morey, 2006a; Mani et al., 2002 and 2006b; Smith et al., 1977). However, the percentage decrease in density depends on the type of biomass. Therefore, a moisture content of 10% (w.b.) is considered as optimal moisture content to obtain high density and durability pellets.

Effect of Grind Size on Pellet Quality

In general, finer grinds produces higher quality pellets since they can readily absorb moisture than large particles, and therefore, undergo a higher degree of conditioning. In addition, finer grinds have higher surface area of contact to form bonds/solid bridges during the compaction processes. Also, large particles are fissure points that cause cracks and fractures in compacts (MacBain, 1966). A reduction in hammer mill screen size from 3.2 to 0.6 mm can result in an increase in pellet densities from 5 to 16% (Kaliyan and Morey, 2006b; Kashaninejad et al., 2011; Mani et al., 2002 and 2004a). However, no significant trend in change in density were observed at geometric mean particles size of 0.6 mm and lower (Kaliyan and Morey 2006b; Mani et al., 2002). The change in pellet density depends on the type of biomass.

This chapter will address various factors that directly or indirectly effect densification of agricultural biomass residue into high quality pellets. The compression and compaction characteristics of ground biomass will be dealt in detail that will provide a comprehensive understanding of the behaviour of biomass as influenced by various factors. The compression studies will explore the affect of independent variables such as biomass, treatment, grind size, and moisture content on pellet density and durability, while compaction studies will study the effect of various machine variables on the pellet quality. In addition, overall specific energy requirements will be established and techno-economic models will be explained.

LIGNOCELLULOSIC COMPOSITION AND HIGHER HEATING VALUES

The experimental lignocellulosic composition of agricultural straw can be determined using the modified NREL LAP method for "Determination of Structural Carbohydrates and Lignin in Biomass" (Table 1) (Adapa et al., 2011; Sluiter et al., 2008). This procedure uses a two-step acid hydrolysis to fractionate the biomass into forms that are more easily quantified. During this process, the lignin fractionates into acid insoluble material and acid soluble material, while the polymeric carbohydrates are hydrolyzed into the monomeric forms, which are soluble in the hydrolysis liquid and subsequently can be measured using HPLC. The Percentage cellulose in the samples can be measured using the percentage glucan content, while the percentage hemicelluloses can be measured by adding the percentage mannose, galactose, xylose and arabinose content in the biomass samples.

Table 1 shows the lignocellulosic composition and higher heating values of non-treated and steam exploded barley, canola, oat and wheat straw samples. In general, the cellulose, hemicelluloses and lignin content of steam exploded straw was higher than non-treated straw. This may be due to other components (soluble lignin, loosely-bound sugars) being washed away during steam explosion, thereby leaving the proportion of insoluble lignin, cellulose and hemicellulose in the resulting dried sample higher than for the non-treated samples (i.e. higher percent of dry mass).

Table 1: Lignocellulosic composition and higher heating values of non-treated and steam exploded agricultural straw (Adapa et al., 2011)

Properties of Biomass	Barley Straw		Canola Straw		Oat Straw		Wheat Straw	
	NT	SE	NT	SE	NT	SE	NT	SE
Composition (% of dry matter)								
Cellulos[e]b	22.7 ± 0.[9]a	25.3 ± 1.8	22.4 ± 0.8	27.5 ± 1.1	25.4 ± 1.0	27.4 ± 2.4	27.1 ± 1.0	29.9 ± 1.4
Hemicellulos[e]c	21.2 ± 0.5	21.0 ± 1.4	16.9 ± 0.5	20.2 ± 0.7	21.7 ± 0.9	18.8 ± 1.2	21.1 ± 0.5	19.7 ± 0.9
Galactose	0.9 ± 0.0	0.7 ± 0.0	1.0 ± 0.0	0.9 ± 0.1	0.8 ± 0.0	0.7 ± 0.0	0.8 ± 0.0	0.9 ± 0.1
Mannose	1.6 ± 0.2	1.5 ± 0.0	2.3 ± 0.1	1.9 ± 0.4	1.4 ± 0.0	1.7 ± 0.1	1.6 ± 0.1	2.8 ± 0.2
Xylose	14.4 ± 0.3	15.3 ± 1.0	11.5 ± 0.5	14.3 ± 0.2	15.1 ± 0.8	13.3 ± 1.0	14.9 ± 0.4	13.5 ± 0.4
Arabinose	4.4 ± 0.2	3.5 ± 0.5	2.0 ± 0.1	3.2 ± 0.0	4.4 ± 0.2	3.1 ± 0.2	3.9 ± 0.1	2.6 ± 0.2
Total Ligni[n]d	21.0 ± 0.6	21.6 ± 0.6	19.6 ± 0.6	22.3 ± 0.2	19.5 ± 0.6	23.7 ± 0.2	22.5 ± 0.7	24.2 ± 0.3
Soluble Lignin	1.6 ± 0.1	1.4 ± 0.1	1.6 ± 0.1	1.2 ± 0.1	1.5 ± 0.1	1.3 ± 0.1	1.4 ± 0.0	1.0 ± 0.1
Insoluble Lignin	19.4 ± 0.6	20.2 ± 0.6	18.0 ± 0.6	21.1 ± 0.1	17.9 ± 0.7	22.4 ± 0.1	21.0 ± 0.7	23.3 ± 0.4
Higher Heating Values (MJ/kg of dry matter)								
HHV (MJ/kg)	16.4±0.3‡†	17.4±0.1	16.7±0.3	18.3±0.0	16.4±0.1	17.8±0.0	17.0±0.2	17.8±0.0

DM – Dry Matter; NT – Non-Treated; SE – Steam Exploded; [a] Average and standard deviation of 3 replicates at 95% confidence interval; [b]%Cellulose = %glucan; [c]%Hemicellulose = %(mannose + galactose + xylose + arabinose);[d]%Total Lignin = %(soluble lignin + insoluble lignin); HHV – Higher Heating Values (measured using Parr 1281 Bomb Calorimeter); ‡3 replicates; † 95% confidence interval

The calorific (heating) value of biomass feedstocks are indicative of the energy they possess as potential fuels. The gross calorific value (higher heating value, HHV) and the net calorific value (lower heating value, LHV) at constant pressure measures the enthalpy change of combustion with and without water condensed, respectively (Demirba , 2007). A bomb calorimeter can be used to determine the HHV of the non-treated and steam exploded straw in MJ/kg. In addition, the ASTM StandardD5865-03 (ASTM, 2003) test method for gross calorific value of coal and coke, can be used as a guideline for heating value testing (Table 1).

Cellulose, hemicelluloses and lignin are major components of a plant biomass. Therefore, a change in composition could potentially lead to change in HHV of the biomass (Adapa et al., 2010a). The Net combined percentage change of cellulose, hemicelluloses and lignin in steam exploded barley, canola, oat and wheat straw is 5%, 19%, 5% and 4% higher than non-treated straw, respectively. As a result, the average HHV of steam exploded barley, canola, oat and wheat straw was 6%, 10%, 9% and 5% higher than non-treated straw, respectively (Table 1).

LAB-SCALE PELLETING OF AGRICULTURAL BIOMASS

Compression Test

A compression apparatus having a close fit plunger die assembly can be used to make a single compact in one stroke of the plunger from ground straw samples (Adapa et al., 2006 and 2010a; Mani et al., 2004). The compression test should be performed to study the effect of independent variables such as biomass, treatment, grind size, and moisture content on pellet density and durability. In order to simulate frictional heating during commercial pelleting operation, the compression die should be maintained at pre-heat temperatures of 75 to 100°C (Adapa et al., 2006; Kaliyan and Morey, 2009;Mani et al., 2006). Different levels of pre-set compressive forces can be applied using the Instron testing machine. Typical pre-set loads in the range of

31.0 to 150.0 MPa are applied to make pellets.Figure 1 represents the photographs of pellets made from barley, canola, oat and wheat straw grinds from hammer mill screen sizes of 3.2, 1.6 and 0.8 mm (Adapa et al., 2010a).

Single-Pellet Density

The density of pellet is calculated from the mass and volume (measuring the length and diameter) of compacts. In general, the density of pellets from agricultural straw significantly increases with an increase in applied pressure at any specific hammer mill screen size. An increase in pressure results in plastic deformation of ground particles and consequently leads to pellets that have densities closer to their respective particle densities (Adapa et al., 2010a; Kaliyan and Morey, 2009; Mani et al., 2004). The Application of pre-treatment has been observed to significantly increase the pellet density since pre-treated straw has lower geometric particle diameters and significantly higher particle densities (Adapa et al., 2010a; Kashaninejad and Tabil, 2011). Usually, it has been reported that an increase in moisture content from 10% and up results in a significant decrease in pellet quality (Hill and Pulkinen, 1988; Li and Liu, 2000; Obernberger and Thek, 2004; Shaw and Tabil, 2007). In general, a decrease in hammer mill screen size results in an increase in pellet density (Adapa et al., 2010a; Kaliyan and Morey, 2009; Kashaninejad et al., 2011; Mani et al., 2004). A comprehensive literature on various single-pellet compression test data is provided in Table 2.

Adapa et al. (2010a) reported that the type of agricultural biomass did not have any significant effect on pellet density, while steam explosion pre-treatment, applied pressure and screen size had significant effects. In addition, correlation for pellet density with applied pressure and hammer mill screen size having highest R^2 values were developed (Table 3). Similarly, Kaliyan and Morey (2009) indicated that the pellet density of corn stover or switchgrass briquettes was significantly affected by pressure, particle size, moisture content and preheating temperature. Kashaninejad and Tabil (2011) also indicated that the pellets made from microwave-chemical pretreated biomass grinds had a significantly higher density and tensile strength than the untreated or samples pretreated by microwave-distilled water.

The densities of pellets should also be measured after a storage period of one week to one month to ascertain its dimensional stability, and associated handling and storage costs (Adapa et al., 2010b;Kaliyan and Morey, 2009). Adapa et al. (2010b) reported that a reduction in pellet density is usually expected due to relaxation of grinds in the pellet after release of pressure. They have observed that the relaxation was higher for larger hammer mill screen sizes and lower applied pressures. In some cases, the average reduction in density was negative giving the impression that pellet density actually increased during storage period. However, these negative values are primarily due to higher standard deviations in pellet density measurements. Therefore, from a practical manufacturing point of view, these values should be considered as a zero percent change in pellet density (Adapa et al., 2010b).

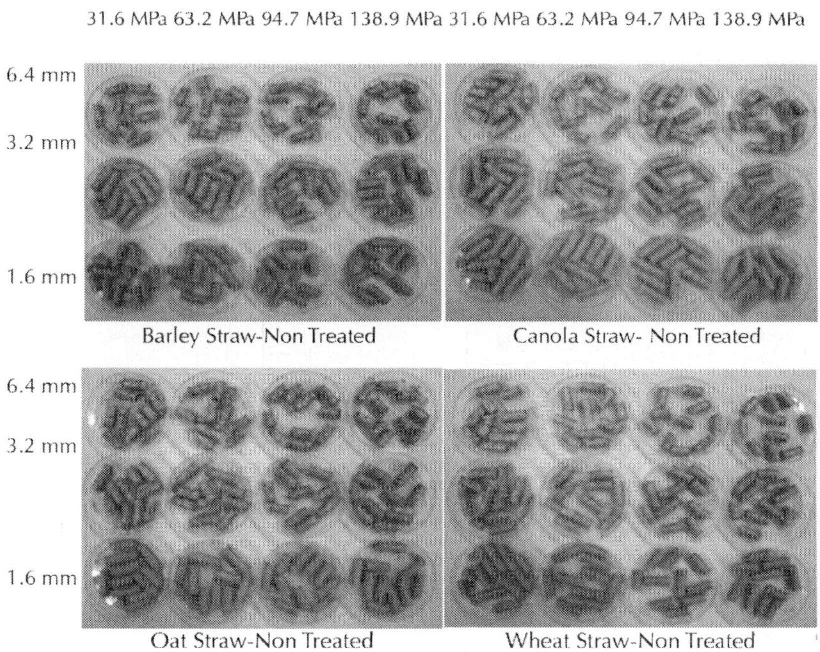

Figure 1: Photograph of pellets made from barley, canola, oat and wheat straw grind from hammer mill screen sizes of 3.2, 1.6, and 0.8 mm.

Table 2: Comprehensive literature review on single-pellet compression tests for agricultural biomass as feedstock for biofuel

Biomass	**Independent Variables**	**Maximum Pellet Density**	**Reference**
Barley, Canola, Oat and Wheat Straw	Hammer Mill Screen Size: 3.2, 1.6 and 0.8 mm Applied Pressure: 31.6, 63.2, 94.7 and 138.9 MPa Moisture Content: 10% (w.b.) Treatment: Non-Treated (NT) and Steam Exploded (SE)	Barley-NT 1003 kg/m3 Barley-SE 1169 kg/m3 Canola-NT 1035 kg/m3 Canola-SE 1165 kg/m3 Oat-NT 1024 kg/m3 Oat-SE 1165 kg/m3 Wheat-NT 1009 kg/m3 Wheat-SE 1180 kg/m3	Adapa et al., 2010a
Corn Stover and Switchgrass	Hammer Mill Screen Size: 3.0 and 4.6 mm Applied Pressure: 100 and 150 MPa Pre-Heat Temperature: 25, 75 and 100°C Moisture Content: 10 and 15% (w.b.)	Corn Stover 1197 kg/m^3 Switchgrass 1098 kg/^{m}3	Kaliyan and Morey, 2009
Barley and wheat straw	Hammer Mill Screen Size: 1.6 mm Applied Pressure: 126.3 MPa Moisture Content: 12% (w.b.) Treatment: Non-Treated (NT), Microwave Pretreated (MT) and Microwave-Chemical Pretreated (MCT)	Barley-NT 995 kg/m3 Barley-MT 984 kg/m3 Barley-MCT 1440 kg/m3 Wheat -NT 950 kg/m3 Wheat -MT 1032 kg/m3 Wheat -MCT 1431 kg/m3	Kashaninejad and Tabil, 2011
Barley and Wheat Straw, Corn Stover, and Switchgrass	Hammer Mill Screen Size: 3.2, 1.6 and 0.8 mm Applied Pressure: 31.6, 63.2, 94.7, 126.3 and 138.9 MPa Moisture Content: 6.22 to 8.30% (w.b.)	Barley Straw 1245 kg/m^3 Wheat Straw 1344 kg/m3 Corn Stover 1399 kg/m3 Switchgrass 1173 kg/m3	Mani et al., 2004

Poplar Wood and Wheat Straw	Hammer Mill Screen Size: 3.2 and 0.8 mm Applied Pressure: 31.6, 63.2, 94.7 and 126.3 MPa Moisture Content: 9 and 15% (w.b.) Pre-Heat Die Temperature: 70 and 100°C Treatment: Non-Treated (NT) and Steam Exploded (SE)	Poplar-NT 1100 kg/m^3 Poplar-SE 1341 kg/m^3 Wheat-NT 1005 kg/m^3 Wheat-SE 1324 kg/^{m}3	Shaw, 2008

Table 3: Correlation for pellet density (, kg/m^3) with applied pressure (P, MPa) and hammer mill screen size (S, mm) for non-treated and steam exploded straw grinds

Independent Variables and Interactions	**Estimated Coefficients**	**R^{2}value**	**Coefficient of Variation**	**Root Mean Square Error**
BarleyNT=587.19+6.29(P)−0.025(P)2				
InterceptPP*P	587.196.29-0.025	0.91	2.99	27.59
BarleySE=727.38+7.41(P)−0.03(P)2−0.08(P×S)+2.84(S)2				
InterceptPP*PP*SS*S	727.387.41-0.03-0.082.84	0.89	3.30	34.89
CanolaNT=545.50+7.41(P)−0.03(P)2−0.11(P×S)				
InterceptPP*PP*S	545.507.41-0.03-0.11	0.89	3.49	32.62
CanolaSE=738.77+7.54(P)−0.03(P)2+0.11(P×S)−28.84(S)				
InterceptPP*PP*SS	738.777.54-0.030.11-28.84	0.92	3.20	22.55
OatNT=666.24+5.79(P)−0.02(P)2−0.12(P×S)+3.54(S)2				
InterceptPP*PP*SS*S	666.245.79-0.02-0.123.54	0.87	3.13	29.00
OatSE=818.41+6.56(P)−0.03(P)2+0.12(P×S)−32.48(S)				
InterceptPP*PP*SS	818.416.56-0.030.12-32.48	0.92	2.66	28.37
WheatNT=700.76+5.99(P)−0.02(P)2−36.06(S)+3.48(S)2				
InterceptPP*PSS*S	700.765.99-0.02-36.063.48	0.90	3.01	27.91
WheatSE=717.60+6.77(P)−0.03(P)2				
InterceptPP*P	717.606.77-0.03	0.91	2.95	31.55
Note: – Density, kg/m^3; NT – Non-Treated; SE – Steam Exploded; P – Pressure, MPa; S – Hammer Mill Screen Size, mm				

Durability

Durability represents the measure of shear and impact forces that a pellet could withstand during handling, storing and transportation process. The durability of pellets is usually measured following the ASABE Standard S269 (ASABE, 2007), which require about 50-100 g of pellets/ compacts. However, due to the limited number of pellets obtained during single-pellet compression test, it is not feasible to use this method. Instead, the durability of pellets can be measured by following the drop test method (Al-Widyan and Al-Jalil, 2001; Khankari et al., 1989; Sah et al., 1980; Shrivastava et al., 1989), where a single pellet is dropped from a 1.85 m height on a metal plate. The larger intact portion of the mass retained is expressed as the percentage of the initial weight.

Adapa et al. (2010) reported that the type of agricultural biomass, steam explosion pretreatment, applied pressure and screen size all had significant effect on pellet durability. Statistically, no significant correlation (R^2 values) was obtained for change in durability with applied pressure and hammer mill screen sizes. In general, pellet durability increases with an increase in applied pressure and grind size, and application of pre-treatment. Similarly, Kaliyan and Morey (2009) indicated that the durability of corn stover or switchgrass briquettes was significantly affected by pressure, moisture content and preheating temperature, while particle size did not have any significant effect.Kashaninejad et al. (2011) also reported the mean durability of pellets made of giant wild rye and mixed forage increased from 63.08 to 89.26% and from 61.47 to 89.21%, respectively when the hammer mill screen size increased from 0.8 to 3.2 mm. This could be primarily due to mechanical interlocking of relatively long fibers at higher grind sizes. They also indicated that at any specific compressive load, the pellet durability of biomass grinds with 12% moisture content was significantly higher than samples with 9 and 15% and demonstrates the moisture contents above or below 12% would lead to lower quality pellets.

Specific Energy for Compaction and Extrusion of Pellet

During the compression and extrusion processes of individual biomass compacts, the force-displacement data is recorded and can be used to calculate the specific compression and extrusion energies following the methodology reported by Adapa et al. (2006) and Mani et al. (2006). The area under the force-displacement curve can be integrated using the trapezoid rule (Cheney and Kincaid, 1980); when combined with the pellet mass, the specific energy values in MJ/t can be calculated.

During single-pellet compression and extrusion, the pellets are prepared by densifying material against a base plate (representing the specific energy required to overcome friction within the straw grinds) as opposed to commercial operation where compacts are formed due to back-pressure effect in the die. Therefore, the specific energy required to extrude the compact should be included, which will closely emulate the specific energy required to overcome the friction between the ground compressed biomass and the die. Mani et al. (2006) have indicated that the extrusion (frictional) energy required to overcome the skin friction was roughly half of the total energy (12-30 MJ/t) for corn stover. Mewes (1959) showed that roughly 40% of the total applied energy was used to compress the materials (straw and hay) and the remaining 60% was used to overcome friction. Faborode and O'Callaghan (1987) studied the energy requirement for compression of fibrous agricultural materials. They reported that chopped barley straw at 8.3% (wb) moisture content consumed 28-31 MJ/t of energy, while un-chopped material consumed 18-27 MJ/t. Shaw (2008) reported that between 95 and 99% of the total specific energy was required to compress the grinds, whereas between 1 and 5% of the total specific energy was required to extrude the compact in single pellet tests. Shaw (2008) also reported that the mean values of specific compression energy ranged from 7.2 (pretreated wheat straw using steam explosion) to 39.1 MJ/t (wheat straw). Kashaninejad and Tabil (2011) indicated that microwave-distilled water and microwave-NaOH pre-treatments significantly increased the specific energy required for compression of wheat straw grinds so that it increased from 16.60 MJ/t to as high as 29.04 and 27.84 MJ/t after pre-treatment by microwave-distilled water and microwave-NaOH, respectively. They also reported less specific energy was required to compress wheat straw pre-treated by combination of microwave and $Ca(OH)_2$. More specific energy was required to eject the pre-treated wheat straw grinds than the untreated wheat straw grinds and it increased from 3.20 MJ/t to 23.08 after pre-treatment by microwave-NaOH. Data analysis showed that the total energy required for compression and ejection of wheat straw grinds pre-treated by microwave-distilled water or microwave-alkaline was higher than untreated samples.

Adapa et al. (2010b) reported that the type of agricultural biomass, steam explosion pretreatment, applied pressure and screen size all had significant effect on specific energy required to form a pellet. In addition, they have developed correlations for specific energy with applied pressure and hammer mill screen size having highest R^2 values for barley, canola, oat and wheat straw (Table 4). In general, the total and compression specific energy for compaction of non-treated and steam exploded barley, canola, oat and wheat straw at any particular hammer mill screen size significantly increased with an increase in applied pressure and significantly decreased with a decrease in hammer mill screen size.

Adapa et al. (2010b) also reported that the specific energy values obtained from the single-pellet compression tests should be used to compare the densification variables. However, these values may not have practical applications since the energy consumed by commercial densification machines / pilot-scale pellet mills may be higher.

COMPRESSION CHARACTERISTICS OF BIOMASS

Compression Mechanism

The compression characteristics of ground agricultural biomass vary under various applied pressures. It is important to understand the fundamental mechanism of the biomass compression process, which is required in the design of energy efficient compaction equipment to mitigate the cost of production and enhance the quality of the product (Mani et al., 2004). To a great extent, the strength of manufactured pellets depends on the physical forces that bond the particles together

(Tabil and Sokhansanj, 1996). These physical forces come in three different forms during pelleting operations:

1. thermal;
2. mechanical; and
3. atomic forces (Adapa et al., 2002).

Pellets are formed by subjecting the biomass grinds to high pressures, wherein the particles are forced to agglomerate. It is generally accepted that the compression process is categorized in several distinct stages and difficult to let one simple monovariate equation to cover the entire densification region (Sonnergaard, 2001). Compression of grinds is usually achieved in three stages (Holman, 1991). In the first stage, particles rearrange themselves under low pressure to form close packing. The particles retain most of their original properties, although energy is dissipated due to inter-particle and particle-to-wall friction. During the second stage, elastic and plastic deformation of particles occurs, allowing them to flow into smaller void spaces, thus increasing inter-particle surface contact area and as a result, bonding forces like van der Waal forces become effective (Rumpf, 1962; Sastry and Fuerstenau, 1973;Pietsch, 1997). Brittle particles may fracture under stress, leading to mechanical interlocking (Gray, 1968). Finally, under high pressure the second stage of compression continues until the particle density of grinds has been reached. During this phase, the particles may reach their melting point and form very strong solid bridges upon cooling (Ghebre-Sellassie, 1989). Figure 2 shows the deformation mechanisms of ground particles under compression (Comoglu, 2007; Denny, 2002).

Table 4: Correlation for pellet density (, kg/m^3) with applied pressure (P, MPa) and hammer mill screen size (S, mm) for non-treated and stem exploded straw grinds

Independent Variables and Interactions	**Estimated Coefficients**	**R^2** value	**Coefficient of Variation**	**Root Mean Square Error**
BarleyNT=587.19+6.29(P)−0.025(P)2				
InterceptPP*P	587.196.29-0.025	0.91	2.99	27.59
BarleySE=727.38+7.41(P)−0.03(P)2−0.08(P×S)+2.84(S)2				

InterceptPP*PP*SS*S	727.387.41-0.03-0.082.84	0.89	3.30	34.89
CanolaNT=545.50+7.41(P)−0.03(P)2−0.11(P×S)				
InterceptPP*PP*S	545.507.41-0.03-0.11	0.89	3.49	32.62
CanolaSE=738.77+7.54(P)−0.03(P)2+0.11(P×S)−28.84(S)				
InterceptPP*PP*SS	738.777.54-0.030.11-28.84	0.92	3.20	22.55
OatNT=666.24+5.79(P)−0.02(P)2−0.12(P×S)+3.54(S)2				
InterceptPP*PP*SS*S	666.245.79-0.02-0.123.54	0.87	3.13	29.00
OatSE=818.41+6.56(P)−0.03(P)2+0.12(P×S)−32.48(S)				
InterceptPP*PP*SS	818.416.56-0.030.12-32.48	0.92	2.66	28.37
WheatNT=700.76+5.99(P)−0.02(P)2−36.06(S)+3.48(S)2				
InterceptPP*PSS*S	700.765.99-0.02-36.063.48	0.90	3.01	27.91
WheatSE=717.60+6.77(P)−0.03(P)2				
InterceptPP*P	717.606.77-0.03	0.91	2.95	31.55
Note: – Density, kg/m^3; NT – Non-Treated; SE – Steam Exploded; P – Pressure, MPa; S – Hammer Mill Screen Size, mm				

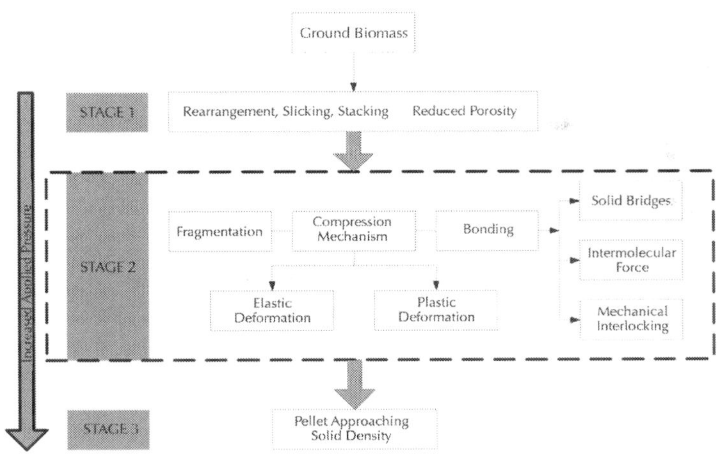

Figure 2: The deformation mechanisms of ground particles under compression (Comoglu, 2007; Denny, 2002).

Compression Characteristics Models

Densification or compaction of agricultural biomass grinds into pellets is an essential process towards production of biofuels. Ground biomass particles behave differently under different applied pressures (Adapa et al., 2002 and 2009a). Therefore, it is important to investigate the change in compact density and volume with pressures. One of the main purposes of fitting experimental data to an equation is usually to develop linear plots in order to make comparisons easier between different sets of data (Comoglu, 2007). A majority of compression models applied to biomass materials have been discussed and reviewed in detail by Adapa et al. (2002 and 2009a), Denny (2002) and Mani et al. (2003). Adapa et al. (2009a) reported that Kawakita and Ludde (1971), Cooper and Eaton (1962) and Jones (1960) models provided the best compression and deformation characteristics of agricultural biomass.

Jones Model

Jones (1960) expressed the density-pressure data of compacted powder in the form of equation 1.

$$\ln(\ )=m\ln(P)+b \qquad (1)$$

where,   is bulk density of compact powder mixture, kg/m^3, P is applied compressive pressure, MPa;m and b are model constants.

The constants b and m are determined from the intercept and slope, respectively, of the extrapolated linear region of the plot of $ln(\ )$ vs $ln(P)$. The constant m has been shown to be equal to the reciprocal of the mean yield pressure required to induce plastic deformation (York and Pilpel, 1973). A large $m$value (low yield pressure) indicates the onset of plastic deformation at relatively low pressure, thus, the material is more compressible.

Cooper-Eaton Model

Cooper and Eaton (1962) studied the compaction behavior of four ceramic powders. In each case it, was assumed that compression is attained by two nearly independent probabilistic processes, namely, the filling of voids having equal size as particles and filling of voids smaller than particles. Based on these assumptions, the following equation (2) was given:

VO−VVO−VS=a1e−k1P+a2e−k2P (2)

where, $V\ O$= volume of compact at zero pressure, m^3; V = volume of compact at pressure P, m^3; $V\ S$= void free solid material volume, m^3; $a\ 1$, $a\ 2$, $k\ 1$, and $k\ 2$ = Cooper-Eaton model constants.

The difficulty in practical use of equation (2) is the assignment of some physical significance to the constant parameters. In addition, another drawback of this model is its applicability to only one-component system (Comoglu, 2007).

Kawakita-Ludde Model

Kawakita and Ludde (1971) performed compression experiments and proposed an equation for compaction of powders based on observed relationship between pressure and volume (Equation 3).

PC=1ab+Pa (3)

Where,

C=VO−VVO (4)

C = degree of volume reduction or engineering strain; a and b = Kawakita-Ludde model constants related to characteristic of the powder.

The linear relationship between P/C and P allows the constants to be evaluated graphically. This compression equation holds true for soft and low bulk density powders (Denny, 2002; Kawakita and Ludde, 1971), but particular attention must be paid on the measurement of the initial volume of the powder. Any deviations from this expression are sometimes due to fluctuations in the measured value of V_0. The constant a is equal to the values of $C = C\infty$ at infinitely large pressure P.

$$C\alpha = VO - V\alpha VO \quad (5)$$

Where, $V\alpha$= net volume of the powder, m^3.

It has been reported that the constant a is equal to the initial porosity of the sample, while constant *1/b*is related to the failure stress in the case of piston compression (Mani et al., 2004).

Compressibility of Different Biomass

The constant *m* in the Jones (1960) model can provided valuable information about the onset of plastic deformation of the ground agricultural biomass. It has been observed that ground particles obtained from larger hammer mill screen sizes has higher compressibility. In addition, application of pre-treatment also improves the compressibility of the agricultural biomass (Adapa et al., 2010).

The dimensionless coefficients, *a 1* and *a 2* in Cooper and Eaton (1962) model represent the densification of powdered material by particle rearrangement and deformation, respectively. If the sum of coefficients (*a 1* + *a 2*) is less than unity, it is an indication that other process must become operative before complete compaction is achieved. For agricultural biomass grinds, the *a 1* values were higher than *a 2* values, hence the material was primarily densified through the process of particle rearrangement. Occasionally, the sum of coefficients (*a 1* + *a 2*) for agricultural biomass was observed to be above unity. The phenomenon of having sum of coefficient more than unity was also observed by Adapa et al. (2002 and 2009a), and Shivanand and Sprockel (1992), which implies that the densification could not be fully attributed to the two mechanisms of compression as assumed by theCooper and Eaton (1962) model (Adapa et al., 2010a).

In the Kawakita and Ludde (1971) model, constant *a* represents the initial porosity of the sample. It has been reported that the porosity and hammer mill screen sizes (corresponding geometric mean particle diameter) are positively correlated. In addition, porosity increases with application of pre-treatment since organized lignocellulosic structure of biomass disintegrates during this process. The parameter *1/b* in the Kawakita-Ludde model indicates the yield strength or failure stress of the compact. In general, the yield strength has negative correlation with hammer mill screen sizes. Also, application of pre-treatment lowers the yield strength of ground agricultural biomass. Statistically, theKawakita and Ludde (1971) model has been observed to provide accurate representation of the compression and deformation characteristics of agricultural biomass (Adapa et al., 2010a).

PILOT-SCALE PELLETING OF AGRICULTURAL BIOMASS

Pilot-scale densification of biomass is required to demonstrate the feasibility of production of pellets by application of various variables studied during single-pellet experiments. A pilot-scale pellet mill such as CPM CL-5 pellet mill (Figure 3) (California Pellet Mill Co., Crawfordsville, IN) can be used for processing of agricultural straw grinds into pellets. The pellet mill usually consists of a corrugated roller and ring die assembly, which compacts and extrudes the biomass grinds from the inside of a ring-shaped die by pressure applied by rolls where either the die or the roll suspension is rotating. Rolls are mounted close to the die surface, but still leaving room for a compacted feed layer to enter the roll gap. Friction between feed layer and rolls makes the rolls rotate (Larsson et al., 2008). In addition to variables indicated in the single-pellet testing, the quality of pellets also depends on machine variables such as the ring die size (radius), length (thickness, l), ring hole diameter (*d*), *l*/*d*ratio, and the rotational speed of the pellet mill (Adapa et al., 2002; Hill and Pulkinen, 1988; Tabil and Sokhansanj, 1996). A monitoring study of commercial pellets was done by Hill and Pulkinen (1988), on variables such as die geometry, conditioning temperatures, natural moisture of the grind, forage quality, bulk density of the grinds, and the use of binding agents. Similarly, Larsson et al. (2008) studied the effect of raw material moisture content, steam addition, raw material bulk density, and die temperature on production of high quality pellets. Also, Serrano et al. (2011), determined the effect of grind size, moisture content and customization of barley straw by adding pine dust to the mixture (blended pellets).

The feed rate of ground biomass to the pellet mill can be controlled using a vibratory feeder (Figure 3). The feed rate should be optimized according to the pellet mill capacity, which will directly affect the throughput. The pilot-scale pelleting test should be performed for a predefined period and the manufactured pellets should be collected and weighed to determine the pellet mill throughput (kg/h). In addition, the pellet mill energy consumption (kWh) should be recorded in real time using a data logger connected to a computer and should be used to calculate the specific energy (MJ/t) required to manufacture pellets from ground agricultural biomass.

Raw materials causing uneven pellet production have low bulk density compared to other milled biofuel pellet raw materials. Low raw material bulk density will put higher demands on the die feeding system of the pelletizer with greater volume throughput for maintained production level. Larsson et al. (2008) investigated the pre-compaction of reed canary grass as an alternative to avoid low and intermittent production of biofuel pellets. They have observed that the process of pre-compaction can increase the bulk density of raw material from 150 kg/m^3 to 270 kg/m^3, which resulted in the continuous production of pellets at a moisture content of 13.8% (w.b.). Pressurized steam conditioners are used in the feed pellet industry to decrease raw material porosity and to improve pellet hardness/ durability (Thomas et al., 1997). Adapa et al. (2010b) were unable to produce any pellets due to the low bulk density of both non-treated and pre-treated agricultural straw grinds at 10% moisture content (w.b.). Therefore, they have added moisture and oil to increase bulk density of grinds to a level of 17.5% (w.b.) and 10% (by weight), respectively, which resulted in production of pellets. Similar observation was made by Serrano et al. (2011) where they have to increase the grind moisture content in the range of 19-23% (w.b.) to produce pellets in a pellet mill. However, addition of pine sawdust to barley straw resulted in high quality pellets at a lower moisture content of 12% (w.b.).

Testing and, if required improving the durability of pellets is important for the industry to evaluate pellet quality and minimize losses during handling and transportation. The concept is not to add any external binders to enhance pellet quality, but rather activate the natural binders in the agricultural biomass by application of various variables, pre-processing techniques and pre-treatments. Biomass pellets can be customized based on proximate analysis data to make them suitable for direct combustion and thermo-chemical conversion applications. Customization can be achieved by forming composites of different straws to control important variables such as energy and ash content of pellets. Similarly, addition of biomass having good binding characteristics to straw with less cohesive characteristics may enhance particle bonding resulting in durable pellets.

Adapa et al. (2010b) reported pellet mill tests on both non-treated and steam exploded agricultural biomass at different hammer mill screen sizes. They have successfully produced pellets from ground non-treated barley, canola, oat and wheat straw at hammer mill screen sizes of 0.8 and 1.6 mm having moisture content of 17.5% (wb) and flax seed oil of 10% by weight. The non-treated ground straw at 3.2 and 6.4 mm screen size did not produce pellets. Similar pelleting process was followed for ground steam exploded straw. Due to very low bulk density and poor flowability, the steam exploded grinds did not produce pellets at any of the hammer mill screen sizes used in the investigation. However, the customized barley, canola, oat and wheat straw having 25% steam exploded material by weight at 0.8 mm screen size successfully produced pellets. Addition of higher percentage of steam exploded straw and customization at screen sizes of 1.6, 3.2, and 6.4 mm did not produce pellets, which could be due to the fact that adding steam exploded (having very low bulk density) to non-treated straw (having relatively higher bulk density) decreased the overall bulk density and flowability of the grinds, thus hindering the production of pellets in the pilot scale mill. The pilot scale pellet mill in this test is constrained with a small motor (3.7 kW (5 hp)) running it, whereas in a commercial pellet mill, the motors are much bigger and more tolerant to changes in feed bulk density. Shaw et al. (2007) reported similar trends where the quality of wheat straw pellets increased with an increase in moisture content to 15.9% (wb).

Figure 4 shows the photograph of pellets manufactured from barley, canola, oat and wheat straw from non-treated grinds at 0.8 and 1.6 mm screen sizes, and customized straw grinds at 0.8 mm having 25% steam exploded straw by weight (Adapa et al., 2010b).

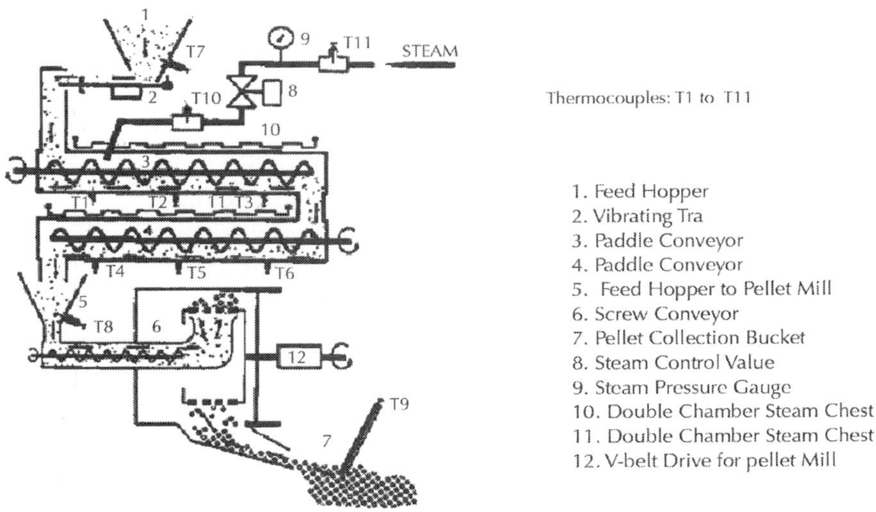

Figure 3: Schematic diagram of CPM CL-5 pellet mill.

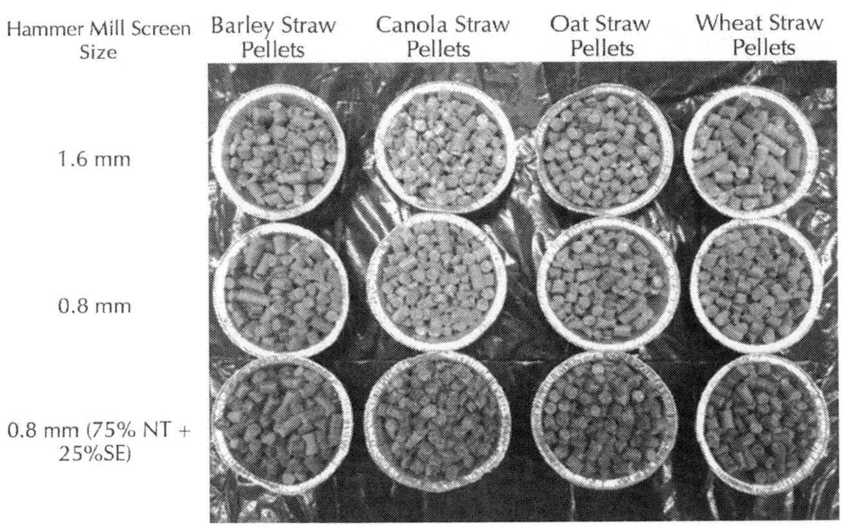

Figure 4: Photograph of pellets manufactured using a pilot scale pellet mill for non-treated (NT) straw at 1.6 and 0.8 mm hammer mill screen size, and customized grinds at 0.8 mm screen size having 25% steam exploded (SE) straw.

Pellet Bulk Density

The mass, length and diameter of individual pellets should be used to determine individual pellet density in kg/m^3. The bulk density of manufactured pellets can be calculated by measuring the mass of pellets filled in a cylindrical container of known volume.

Larsson et al. (2008) reported that the most influential factor for the pellet bulk density was raw material moisture content and showing a negative correlation. Similarly, two other studies have observed that the bulk density of wheat straw, big bluestem grass, corn stover, sorghum stalk and switchgrass decreased with an increase in moisture content (Colley et al., 2006; Theerarattananoon, et al., 2011). Larsson et al. (2008) did not find any correlation between pellet bulk density and die temperature, which contradicts to the observations made by Hill and Pulkinen,1988, and Tabil and Sokhansanj (1996). Serrano et al. (2011) did not observe any significant effect of hammer mill screen size (4 mm and 7 mm) on pellet density. However, pellet density decreased with an increase in moisture content.

Adapa et al. (2010b) reported pellet density obtained from non-treated straw samples at 1.6 and 0.8 mm, and customized sample having 25% steam exploded straw at 0.8 mm screen size (Table 5). In general, pellet density increased with a decrease in screen size from 1.6 to 0.8 mm. However, no significant differences in density values were observed for non-treated samples at 0.8 mm and customized samples, except for canola and oat straw. This could be due to large fluctuation in individual pellet density values. All of the pellet density values reached near individual biomass particle densities at respective grind sizes (Adapa et al., 2010b).

Bulk density of pellets from barley, canola, oat and wheat straw showed significant difference with grind size and customization, except for wheat straw pellets at 0.8 mm for non-treated and customized samples (Table 5). In general, the average pellet bulk densities obtained for customized straw samples were higher (except for barley straw), which is consistent with increase in particle densities. The bulk densities of pellets manufactured were higher than the minimum design value of 650 kg/m^3 suggested by Obernberger and Thek (2004) for wood pellet producers.

Pellet Durability

The durability of pellets can be measured following the ASABE Standard S269 (2007). The method states that 100 g of pellet sample should be weighed and placed in a dust-tight enclosure/ chamber, and tumbled for 10 min at 50 r/min. A 5.70 mm sieve should used to determine the fines produced by the pellets during the tumbling process. The mass of pellets left on the sieve, as percentage of the total mass of pellet sample used during the test, was considered as the durability of the pellets.

Hill and Pulkinen (1988) reported an increase in pellet durability by about 30 to 35% with an increase in pellet temperature from 60 to 104°C. A die length-to-diameter ratio (*l/d*) of 8 to 10 was also reported to be ideal for making high quality pellets. Similarly, Tabil and Sokhansanj (1996) conducted a study for improving the physical quality of alfalfa pellets by controlling and optimizing the manufacturing process. The process conditions investigated were steam conditioning temperatures, die geometry (length to diameter or *l/d* ratio), hammer mill screen sizes used in grinding dry chops, and die speed. They reported that a higher conditioning temperature (95°C) resulted in improved durability of processed pellets. The durability of samples was generally better using the smaller die (higher *l/d*ratio). The hammer mill screen size did not show any affect on pellet durability. Finally, they reported that high durable pellets are obtained at low die speed (250 rpm). Theerarattananoon et al. (2011) also observed that an increase in length / thickness of die resulted in significant increase in durability of biomass pellets.

Larsson et al. (2008) and Serrano et al. (2011) reported that the most influential factor affecting pellet durability was raw material moisture content and showing a positive correlation. The maximum durability was obtained at moisture content of 14.9% without steam addition, and at 13.7% (w.b.) with steam addition of 2.6% (Larsson et al., 2008). Serrano et al. (2011) found that the highest mechanical durability reached for barley straw pellets was 95.5% at moisture content of 19-23% (w.b.), while no pellets were formed below the 19% moisture content. In addition, they observed that the durability of barley straw pellets increased with addition of pine sawdust at 2, 7 and 12% bu weight. However, the percentage increase in pine sawdust did not have significant effect on durability. Colley et al. (2006) observed highest durability of 95.9% for pellets from switchgrass at a moisture content of 8.6% (w.b.). The moisture content in the range of 9-14%, 9-11% and 14-16% (d.b.) did not have any significant effect on maximum durability of 96.8%, 96.8% and 89.5% for pellets from wheat straw and corn stover, big bluestem grass, and sorghum stalk, respectively (Theerarattananoon, et al., 2011); however, further increasing the moisture content reduced pellet durability for respective agricultural biomass. Also, the durability was observed to be positively correlated to die temperature (Larsson et al., 2008).

Adapa et al. (2010b) reported that the durability of pellets obtained from non-treated straw samples at 1.6 and 0.8 mm, and customized sample having 25% steam exploded straw at 0.8 mm screen size were significantly different (Table 5). In general, higher durability values were observed for non-treated straw samples at 0.8 mm hammer mill screen size. The durability of pellets significantly increased with a decrease in grind size for non-treated samples from 1.6 to 0.8 mm. However, addition of steam exploded straw to non-treated straw at 0.8 mm screen size resulted in a decrease in durability, except for wheat straw. This could be due to the fact that steam exploded material has lower soluble lignin content and higher cellulose and hemicelluloses content compared to non-treated straw (Table 1). This observation is in contrast to Lam et al. (2008), who reported that the quality (durability) of pellets produced from steam exploded sawdust was 20% higher than non-treated sawdust. Though, it is important to note that high durability values (>80%) were obtained for all pilot scale pelleting tests.

Specific Energy and Energy Balance during Pelleting

The durability of pellets was negatively correlated to pellet mill throughput and was positively correlated to specific energy consumption (Table 5). The specific energy values obtained from pilot scale pellet mill are 10-25 times higher than reported by Mani et al. (2006b) and Adapa et al. (2010aand 2009b) for agricultural straw, using a single pellet Instron testing machine. The higher pellet mill specific energy numbers could be due to higher friction values and practical pelleting conditions, which are closer to industrial operations.

An overall specific energy analysis is desired in order to understand the net amount of energy available for the production of biofuels after postharvest processing and densification of agricultural straw. The specific energy analysis was performed for pilot-scale pelleting of non-treated and customized (75% non-treated + 25% steam exploded) barley, canola, oat and wheat straw at 1.6 and 0.8 mm hammer mill screen sizes (Table 6). The specific energy for grinding of straw at 0.8 mm was calculated using regression equations reported in Adapa et al. (2011b). The specific energy for chopping and grinding of biomass, production of pellets using pellet mill and higher heating values for straw were obtained from experimental data (Adapa et al., 2011b and 2010b). In addition, the specific energy required for operating the chopper, hammer mill and pellet mill were 337, 759 and 429 W, respectively. On average, the operation of biomass chopper required five times more energy than chopping of biomass. On the other hand, the grinding of biomass required on an average three times more energy than operation of hammer mill. Interestingly, almost the same amount of energy was required to operate the pellet mill and production of pellets. Total specific energy required to form pellets increased with a decrease in hammer mill screen size from 1.6 to 0.8 mm, however, the total specific energy for the process decreased for customized straw compared to non-treated straw at 0.8 mm screen size (Table 6). It has been determined that the net specific energy available for production of biofuel is a significant portion of original agricultural biomass energy (92-94%) for all agricultural biomass (Table 6).

Cost and Life Cycle Assessment of Biomass Densification

Sultana et al. (2010) performed a techno-economic analysis and developed a model for a plant that can produce agricultural straw (barley, oat and wheat) pellets for 30 years. They have included the cost of obtaining the straw, transporting straw to the pellet plant, and producing pellets. Costs incurred by the plant for the production of pellets included capital cost, energy cost, labor cost, and consumable cost. The biomass procurement area was determined to estimate the transportation cost. The scale factors for all the equipment related to pellet production were determined based on the data of previous studies (Sultana et al., 2010). To develop the model, minimum, average and maximum yields of wheat, barley and oats straw in Western Canada were considered. They have determined that the cost of pellets does not change much for capacities over 70,000 tonnes per year (cost of production per tonne is $170.89). Therefore, the optimum size is the same for both average and maximum yield cases. In addition, it was observed that the total cost of pellet production is most sensitive to field cost followed by the transportation cost.

Life cycle assessment (LCA) study was performed on wheat straw production system and densification system in the Canadian Prairies using the LCA modelling software tool SimaPro 7.2 to determine the environmental burdens of manufacturing the wheat straw bale and wheat straw pellet (Li et al., 2011). The factors taken into consideration were greenhouse gas emission, acidification, eutrophication, ozone layer depletion, abiotic depletion, human toxicity, and photochemical oxidation. Li et al. (2011) reported that the production of biomass pellet has higher global warming effect than biomass bale, especially in CO_2 and CH_4 emissions from fossil fuel consumption, which is very high in densification system due to machinery usage. It was also reported that the production of wheat straw pellet has higher environmental impact on acidification, eutrophication, human toxicity and other categories than biomass bale. The dominant factors determining most environmental impacts in agricultural system are fertilizer use and production, while machinery use, manufacturing and energy consumption are main contributors to greenhouse gas emission and other environmental burdens in the densification system (Li et al., 2011).

Table 5: Pellet density, durability, throughput and specific energy data for non-treated and steam exploded barley canola, oat and wheat straw at 17.5% moisture content (wb) and 10% flaxseed oil content

Agricultural Biomass	**Hammer Mill Screen Size (mm)**	**Pellet Density (kg/m³)**	**Pellet Bulk Density (kg/m³)**	**Durability (%)**	**Throughput (kg/h)**	**Specific Energy (MJ/t)**
Barley Straw	1.6 (100% NT)	1158±109*†£ aD	665±01‡ aD	91±00‡ aD	4.88	293
	0.8 (100% NT)	1174±46 aD	700±07 bD	93±01 bD	4.21	353
	0.8 (75% NT + 25% SE)	1184±63 aD	714±02 cD	87±01 cD	3.46	301
Canola Straw	1.6 (100% NT)	1023±85 aE	629±01 aE	90±01 aD	3.86	385
	0.8 (100% NT)	1204±43 bDE	720±04 bE	95±00 bE	3.63	440
	0.8 (75% NT + 25% SE)	1144±50 cD	641±01 cE	82±00 cE	5.51	265
Oat Straw	1.6 (100% NT)	1140±63 abD	631±03 aE	89±01 aE	4.48	340
	0.8 (100% NT)	1188±78 aDE	649±02 bF	93±00 bD	3.81	344
	0.8 (75% NT + 25% SE)	1071±101 bE	676±06 cF	89±01 aF	4.03	335
Wheat Straw	1.6 (100% NT)	1163±57 aD	673±02 aF	94±01 aF	5.44	381
	0.8 (100% NT)	1278±136 bE	721±04 bE	95±01 bE	3.81	297
	0.8 (75% NT + 25% SE)	1213±88 abD	722±04 bG	95±00 cG	4.08	342

NT – Non-treated Straw Samples; SE – Steam Exploded Straw Samples; *10 replicates; ‡3 replicates; † 95% confidence interval; £ Student-Neuman-Keuls test at 5% level of significance for same sample biomass at various hammer mill screen sizes (a, b and c); at same hammer mill screen size for different sample biomass (D, E, F and G)

Table 6: Overall specific energy analysis to show net energy available for production of biofuels after postharvest processing and densification of agricultural straw

Treatment	Hammer Mill Screen Size (mm)	Specific Energy (MJ/t)				HHV (MJ/t)	Net Energy (MJ/t)
		Chopping Biomass	Grinding Biomass	Pilot-Scale Pelleting	Total£		
Barley							
NT*	1.6	11.3	90.4	293	924	16400	15476
NT	0.8	11.3	206.6	353	1100	16400	15300
75% NT + 25% SE*	0.8	11.3	189.3	301	1030	16650	15620
Canola							
NT	1.6	7.1	128.5	385	987	16700	15713
NT	0.8	7.1	363.3	440	1277	16700	15423
75% NT + 25% SE	0.8	7.1	341.6	265	1080	17100	16020
Oat							
NT	1.6	9.9	149.5	340	1029	16400	15371
NT	0.8	9.9	253.6	344	1137	16400	15263
75% NT + 25% SE	0.8	9.9	245.2	335	1120	16750	15630
Wheat							
NT	1.6	8.2	153.3	381	1048	17000	15952
NT	0.8	8.2	382.7	297	1194	17000	15806
75% NT + 25% SE	0.8	8.2	332.1	342	1188	17200	16012

*NT – Non-Treated; SE – Steam Exploded£ Total Specific Energy = Specific Energy (Chopping Biomass + Operating Chopper + Grinding Biomass + Operating Hammer Mill + Pilot-Scale Pelleting) Net Energy = HHV – Total

SUMMARY

The densification of biomass into durable compacts is an effective solution to meet the requirement of raw material for biofuel production. The compression characteristics of ground agricultural biomass vary under various applied pressures. It is important to understand the fundamental mechanism of the biomass compression process, which is required to design an energy efficient compaction equipment to mitigate the cost of production and enhance the quality of the product. To a great extent, the strength of manufactured compacts depends on the physical forces that bond the particles together. These physical forces are generated in three different forms during compaction operations:

1. thermal;
2. mechanical; and
3. atomic forces.

To customize and manufacture high quality products that can withstand various forces during transportation and handling, it is essential to predict desirable and dependent quality parameters (density and durability) with respect to various independent variables (pre-treatment, grind size, applied pressure, hold time, die temperature, and moisture content). In addition, specific energy requirements of manufacturing biomass pellets should be established, which can assist in determining the economic viability of densification process.

The density of biomass pellet has been observed to significantly increase with an increase in applied pressure and a decrease in hammer mill screen size. In addition, application of pre-treatment has observed to significantly increase the pellet density since pre-treated straw has lower geometric particle diameters and significantly higher particle densities. Statistically, agricultural biomass did not have any significant effect on pellet density, while steam explosion pre-treatment, applied pressure, moisture content, pre-heat temperature and screen size had significant effect. A negative correlation has been observed between the pellet bulk density and moisture content, while a positive correlation exists between bulk density and pellet mill die temperature. In general, average pellet bulk densities obtained for customized straw samples is higher as a direct result of increase in particle densities.

Agricultural biomass, steam explosion pre-treatment, applied pressure, moisture content, pre-heat temperature and screen size all had significant effect on pellet durability. In general, durability of pellets increases with an increase in applied pressure and grind size, and application of pre-treatment. An increase in pellet mill die temperature, steam conditioning temperature and die thickness resulted in an increase in pellet durability. No specific trend in durability was observed with customization of straw by mixing non-treated and steam exploded straw grinds.

The specific energy required to form a pellet has been significantly affected by the type of agricultural biomass, steam explosion pre-treatment, applied pressure and screen size. The total and compression specific energy for compaction of non-treated and steam exploded barley, canola, oat and wheat straw at any particular hammer mill screen size significantly increased with an increase in applied pressure and significantly decreased with a decrease in hammer mill screen size. Durability of pellets was negatively correlated to pellet mill throughput and was positively correlated to specific energy consumption. An overall energy balance was performed, which showed that a significant portion of original agricultural biomass energy (92-94%) is available for the production of biofuels.

REFERENCES

1. P. K. Adapa, L. G. Tabil, G. J. Schoenau, T. Canam, T. Dumonceaux, 2011aQuantitative Analysis of Lignocellulosic Components of Non-Treated and Steam Exploded Barley, Canola, Oat and Wheat Straw using Fourier Transform Infrared Spectroscopy. The Journal of Agricultural Science and Technology, 5(7), (In-Press).
2. Adapa, P.K., Tabil, L.G., & Schoenau, G.J.2011bGrinding Performance and Physical Properties of Non-Treated and Steam Exploded Barley, Canola, Oat and Wheat Straw. Journal of Biomass and Bioenergy,doi:10.1016/j.biombioe.2010.10.004, 35,549 561
3. P. K. Adapa, L. G. Tabil, G. J. Schoenau, 2010a Compression Characteristics of Non-Treated and Steam Exploded Barley, Canola, Oat and Wheat Straw Grinds.ASABE Journal of Applied Engineering in Agriculture 617 632

4. P. K. Adapa, L. G. Tabil, G. J. Schoenau, 2010b Pelleting Characteristics of Selected Biomass with and without Steam Explosion Pretreatment International Journal of Agricultural and Biological Engineering 62 79
5. P. K. Adapa, L. G. Tabil, G. J. Schoenau, 2009aCompression Characteristics of Selected Ground Agricultural Biomass. Agricultural Engineering International: the CIGR Ejournal, Manuscript 1347, XI(June), 1 19
6. Adapa, P.K., Tabil, L.G., & Schoenau, G.J.2009bCompaction Characteristics of Barley, Canola, Oat and Wheat Straw. Biosystems Engineeringdoi:10.1016/j.biosystemseng.2009.06.022,104,335 344
7. P. K. Adapa, G. J. Schoenau, L. G. Tabil, E. A. Arinze, A. Singh, A. K. Dalai, 2007Customized and value-added high quality alfalfa products- a new concept. Agricultural Engineering International: the CIGR Ejournal, Manuscript FP 07 003, IX(June), 1 28
8. P. K. Adapa, A. Singh, G. J. Schoenau, L. G. Tabil, 2006Pelleting Characteristics of Fractionated Alfalfa Grinds- Hardness Models. International Journal of Powder Handling and Processing, 18(5), 294 299
9. P. K. Adapa, L. G. Tabil, G. J. Schoenau, B. Crerar, S. Sokhansanj, 2002Compression Characteristics of Fractionated Alfalfa Grinds. Powder Handling and Processing, 14(4), 252 259
10. M. I. Al-Widyan, H. F. Al-Jalil, 2001Stress-density relationship and energy requirement of compressed only cake. Applied Engineering in Agriculture, 17(6), 749 753
11. ASABE Standards 269.4. 2007Cubes, Pellets and Crumbles- Definitions and methods for determining density, durability and moisture content. In: ASABE Standards. St. Joseph, MI.: American Society of Agricultural and Biological Engineers.
12. A. S. T. M. A. S. T. M. D5865-0, 2003Standard test method for gross calorific value of coal and coke. In: Annual book of ASTM standards, 05West Conshohocken, PA.: American Society for Testing and Materials.
13. J. L. Bowyer, V. E. Stockmann, 2001Agricultural residues: An exciting bio-based raw material for the global panel industry. Forest Products Journal, 51(1), 10 21

14. Campbell, C.A., Zentner, R.P., Gameda, S., Blomert, B. & Wall, D.D.2002Production of annual crops on the canadian prairies: trends during 1976- 1998. Canadian Journal of Soil Science, 8245 57
15. W. J. Chancellor, 1962Formation of hay wafers with impact loads. Agricultural Engineering, 43(3), 136 138
16. W. Cheney, D. Kincaid, 1980 Numerical mathematics and computingMonterey, CA: Brooks/Cole Publishing Company.
17. Colley, Z., Fasina, O.O., Bransby, D., & Lee, Y.Y. (2006). pp. 1845-1851 Moisture effect on the physical characteristics of switchgrass pellets. Transactions of ASABE, 49,1845 1851
18. T. Comoglu, 2007An Overview of Compaction Equations. Journal of Faculty of Pharmacy, Ankara 36(2), 123 133
19. A. R. Cooper, L. E. Eaton, 1962 Compaction Behavior of Several Ceramic Powders Journal of the American Ceramic Society 97 101
20. Demirbaş, 2007Effect of moisture and hydrogen content on the heating value of fuels. Energy Sources Part A, 29, 649 655
21. Demirbaş, A. Sahin, 1998Evaluation of biomass residue 1. Briquetting waste paper and wheat straw mixtures. Fuel Processing Technology, 55 175 183
22. P. J. Denny, 2002 Compaction Equations: A Comparison of the Heckel and Kawakita Equations Powder Technology 162 172
23. M. O. Faborode, J. R. O'Callaghan, 1987 Optimizing the compression/ briquetting of fibrous agricultural materials Journal of Agricultural Engineering Research 245 262
24. FAO, 2011Gasification Fuels. Food and Agriculture Organization of the United Nations. Website accessed on March 15, Web Address: http://www.fao.org/docrep/t0512e/t0512e0b.htm
25. Ghebre-Sellassie, 1989Mechanism of Pellet Formation and Growth. Pharmaceutical Pelletization Technology, ed. I. Ghebre-Sellassie, 123 143New York, NY: Marcel Dekker Inc.
26. W. A. Gray, 1968Compaction after Deposition. In The Packing of Solid Particles, 89 107New York, NY: Marcel Dekker Inc.
27. P. D. Grover, S. K. Mishra, 1996 Biomass Briquetting Technology and Practices.Food and Agriculture Organization (FAO), UN, Document 46

28. A. S. Gustafson, W. L. Kjelgaard, 1963 Hay pellet geometry and stability. AgriculturalEngineering, 44(8), 442 445

29. A. Hill, D. A. Pulkinen, 1988 A study of the factors affecting pellet durability and pelleting efficiency in the production of dehydrated alfalfa pellets. Saskatchewan, Canada: Saskatchewan Dehydrators Association.

30. A. A. Holley, 1983The densification of biomass by roll briquetting. Proceedings of the Institute for Briquetting and Agglomeration (IBA), 18, 95 102

31. L. E. Holman, 1991 The Compaction Behaviour of Particulate Materials: An Elucidation based on Percolation Theory Powder Technology 265

32. W. D. Jones, 1960Fundamental Principles of Powder Metallurgy; London UK; Edward Arnold Publishers Ltd.: 242 370

33. N. Kaliyan, R. V. Morey, 2006a Factors affecting strength and durability of densified products. ASABEAnnual International Meeting, American Society of Agricultural and Biological Engineers, Portland, Oregon July 9 12Paper 066077Niles Road, St. Joseph, MI 49085-9659 USA.

34. N. Kaliyan, R. V. Morey, 2006b Densification characteristics of corn stover and switchgrassASABE Annual International Meeting, American Society of Agricultural and Biological Engineers, Portland, Oregon July 9 12Paper 066174Niles Road, St. Joseph, MI 49085-9659 USA.

35. N. Kaliyan, R. V. Morey, 2009 Densification Characteristics of Corn Stover and Switchgrass Transactions of the ASABE 907 920

36. M. Kashaninejad, L. G. Tabil, 2011 Effect of microwave-chemical pre-treatment on compression characteristics of biomass grinds Biosystems Engineering 36 45

37. M. Kashaninejad, L. G. Tabil, X. Tabil, 2011Compression characteristics of giant wild rye and mixed forage grinds, Renewable Energy, Under review.

38. Kawakita, K. Ludde, H. , 1971 Some Considerations on Powder Compression Equations Powder Technology 61 68

39. K. Khankari, M. Shrivastava, R. V. Morey, 1989Densification characteristics of rice hulls. ASAE Paper 89-6093 89 6093St. Joseph, Mich., ASABE.

40. P. S. Lam, S. Sokhansanj, X. Bi, C. J. Lim, S. Mani, S. Melin, 2008Energetic feasibility of biomass densification with steam explosion pretreatment. CSBE Annual General Meeting; July 14 16Poster CSBE08305Vancouver, BC.
41. S. H. Larsson, M. Thyrel, P. Geladi, T. A. Lestander, 2008High quality biofuel pellet production from pre-compacted low density raw materials. Bioresource Technology, 99, 7176 7182
42. Y. Li, H. Liu, 2000 High-pressure densification of wood residues to form an upgraded fuel Biomass and Bioenergy 177 186
43. R. Liu, H. Yu, Y. Huang, 2005 Structure and morphology of cellulose in wheat straw Cellulose 25 34
44. Bain. R. Mac, 1966Pelleting animal feed. Chicago, IL: American Feed Manufacturing Association.
45. S. Mani, L. G. Tabil, S. Sokhansanj, 2006 Effects of compressive force, particle size and moisture content on mechanical properties of biomass pellets from grasses Biomass and Bioenergy 1420 1426
46. S. Mani, L. G. Tabil, S. Sokhansanj, 2006b Specific energy requirement for compacting corn stover Bioresource Technology 1420 1426
47. S. Mani, L. G. Tabil, S. Sokhansanj, 2004 Evaluation of compaction equations applied to four biomass species Canadian Biosystems Engineering 3
48. S. Mani, L. G. Tabil, S. Sokhansanj, 2003An overview of compaction of biomass grinds. Powder Handling & Process, 15(3),160 168
49. S. Mani, L. G. Tabil, S. Sokhansanj, 2002Compaction behavior of some biomass grinds. AIC Paper 02-305 02 305Saskatoon, Saskatchewan: AIC Meeting, CSAE/SCGR Program.
50. J. Mc Mullen, O. O. Fasina, C. W. Wood, Y. Feng, 2005 Storage and handling characteristics of pellets from poultry litter Applied Engineering in Agriculture 645 651
51. E. Mewes, 1959Verdichtungsgesetzmassigkeiten nach presstopfversuchen (Compression relationships as a result of experiments in pressure chambers). Landtechnische Forschung, 9(3), 68 76

52. Obernberger, G. Thek, 2004Physical characterization and chemical composition of densified biomass fuels with regard to their combustion behavior. Biomass and Bioenergy, 27, 653 669
53. W. Pietsch, 1997 Size Enlargement by AgglomerationIn Handbook of Powder Science and Technology, 2nd edition, eds. M.E. Fayed and L. Otten, 202 377Florence, KY: International Thomson Publishing.
54. A. Rajvanshi, 1986Published as Chapter 4 in the book Alternative Energy in Agriculture, Vol. II, Ed. D. Yogi Goswami, CRC Press, Boca Raton, FL.
55. H. Rumpf, 1962The Strength of Granules and Agglomerates. In Agglomeration, ed. W.A. Knepper, 379 419New York, NY: John Wiley and Sons.
56. P. Sah, B. Singh, U. Agrawal, 1980Compaction behavior of straw. Journal of Agricultural Engineering-India, 18(1), 89 96
57. V. S. Sastry, D. W. Fuerstenau, 1973 Mechanisms of Agglomerate Growth in Green Pelletization Powder Technology 97 105
58. Serrano, E. Monedero, M. Laupuerta, H. Portero, 2011 Effect of Moisture Content, Particle Size and Pine Addition on Quality Parameters of Barley Straw Pellets Fuel Processing Technology 699 706
59. D. Shaw, L. G. Tabil, 2007 Compression and relaxation characteristics of selected biomass grinds.ASAE Annual International Meeting, Minneapolis, MN, June 17 20Paper 076183Niles Road, St. Joseph, MI 49085-9659 USA.
60. D. Shaw, L. G. Tabil, A. Opoku, 2007Pelleting characteristics and related properties of poplar wood and wheat straw grinds. A report prepared for SunOpta Bioprocess Inc., Brampton, ON, Canada, 16 p.
61. D. Shaw, 2008 Feedstock and Process Variables Influencing Biomass DensificationM.Sc. dissertation. Saskatoon, SK, Canada: University of Saskatchewan.
62. Shivanand, O. L. Sprockel, 1992 Compaction Behavior of Cellulose Polymers Powder Technology 177 184
63. Shrivastava, M., Shrivastava, P., & Khankari, K.K.1989Densification characteristics of rice husk under cold and hot compression. In Agricultural Engineering: Proceedings of the 11th International

Congress on Agricultural Engineering, Dublin, 4-8 September, 2441-2443. V.A. Dodd and P. M. Grace, eds. Rotterdam: A.A. Balkema Pub4 8

64. A. Sluiter, B. Hames, R. Ruiz, C. Scarlata, J. Sluiter, D. Templeton, D. Crocker, 2008Determination of Structural Carbohydrates and Lignin in Biomass, A Technical Report: NREL/TP-510 42618National Renewable Energy Laboratory, USA, Available online at: http://www.nrel.gov/biomass/pdfs/42618.pdf
65. E. Smith, S. Probert, R. Stokes, R. Hansford, 1977 The briquetting of wheat straw Journal of Agricultural Engineering Research 105 111
66. S. Sokhansanj, S. Mani, M. Stumborg, R. Samson, J. Fenton, 2006Production and distribution of cereal straw on the canadian pariries. Canadian Biosystems Engineering, 48, 3
67. S. Sokhansanj, S. Mani, X. Bi, P. Zaini, L. G. Tabil, 2005 Binderless pelletization of biomassASAE Annual International Meeting, Tampa Convention Centre, Tampa, Florida July 17 20Paper 056061Niles Road, St. Joseph, MI 49085-9659 USA.
68. M. Sonnergaard, 2001 Investigation of a New Mathematical Model for Compression of Pharmaceutical Powders European Journal of Pharmaceutical Sciences 149 157
69. A. Sultana, A. Kumar, D. Harfield, 2010 Development of Agri-Pellet Production Cost and Optimum Size BioresourceTechnology 5609 5621
70. G. Tabil, S. Sokhansanj, 1996Compression and Compaction Behavior of Alfalfa Grinds Part 1- Compression Behaviour. Journal of Powder Handling and Processing, 8(1), 17 23
71. Temmerman, F. Rabier, P. D. Jensen, H. Hartmann, T. Bohm, 2006 Comparative study of durability test methods for pellets and briquettes Biomass and Bioenergy 964
72. K. Theerarattananoon, F. Xu, J. Wilson, R. Ballard, L. Mc Kinney, S. Staggenborg, P. Vadlani, Z. J. Pei, D. Wang, 2011 Physical Properties of Pellets Made from Sorghum Stalk, Corn Stover, Hwaet Straw and Big Bluestem. Industrial Crops and Products 325 332

73. Thomas, D. J. van Zuilichem, A. F. B. van der Poel, 1997 Physical quality of pelleted animal feed. 2. Contribution of processes and its conditions Animal Feed Science and Technology 173 192
74. L. Xue, E. Mupondwa, S. Panigrahi, L. Tabil, P. Adapa, 2011Life Cycle Assessment of Biomass Densification System in Canadian Prairies. Personal Communication.
75. York, N. Pilpel, 1973The Tensile Strength and Compression Behavior of Lactose, Four Fatty Acids and their Mixture in relation to Tableting. Journal of Pharmacy and Pharmacology, 25, 1 11

Chapter 4

Conversion of Non-HomogeneousBiomass to Ultraclean Syngas and Catalytic Conversion to Ethanol

Stéphane C. Marie-Rose[1], Alexis Lemieux Perinet[1], and Jean-Michel Lavoie[1]

[1] Industrial Research Chair on Cellulosic EthanolDepartment of Chemical and Biotechnological EngineeringUniversité de Sherbrooke, Sherbrooke, Québec,, Canada

INTRODUCTION

Reducing greenhouse gas emissions, rising energy prices and security of supply are reasons that justify the development of biofuels. However, food prices recorded in 2007 and 2008 affected more than 100 million of people that became undernourished worldwide (Rastoin, 2008) The food crisis has been caused by several factors: underinvestment in agriculture, heavy speculation on agricultural commodities and competition of biofuels vs. food. It is estimated that by 2050, it will be essential to increase by 50% the food production to support the 9 billion people living on the planet (Rastoin, 2008).

Recycling the carbon from residual waste to produce biofuels is one of the challenges of this new century. Several companies have been developing technologies that are able to transform residual streams into syngas, which is subsequently converted into alcohols. "Green" ethanol plays an important role in reducing dependency toward petroleum and providing environmental benefit, through its role in the fuel additive market. Ethanol is an oxygenate and also serves as an octane enhancer. The waste-to-syngas approach is an alternative to avoid the controversy food vs. fuel whilst reducing landfills and increasing carbon recuperation. Using this approach, yields of ethanol produced are above 350 liters/dry tonne of feedstock entering the gasifier (Enerkem's technology is taken as example). Residual heat, also a product of the process, is used in the process itself and, as well, it can be used for outside heating or cooling. Enerkem Inc. is moving the technology from bench scale, to pilot, to demo to commercial implementation (a 12,500 kg/h of sorted and biotreated urban waste, is being constructed in Edmonton, Alberta). Economics of the process are favorable at the above commercial capacity, given the modular construction of the plant, reasonable operational costs and a tipping fee for the residue going into the gasifier.

The first part of this chapter will present feedstock preparation, gasification and gas conditioning. The characteristics of the heterogeneous feedstock will determine its performance during gasification for syngas production whose composition has the appropriate H_2/CO ratio for downstream synthesis. The second part of the chapter will be directed at the methanol synthesis in a three-phase reactor using syngas. The third and last part of the chapter will focus on the catalytic steps to convert methanol into bio-ethanol.

SYNTHESIS GAS (SYNGAS) PRODUCTION BY GASIFICATION

Characteristics and Composition of Heterogeneous Wastes as Feedstock

Biomass is defined as an organic material derived from plants or animals that contain potential chemical energy; for example wood, which was the first fire source used by mankind, and which is still used today by population for cooking and heating.

At world scale, biomass is now the fourth largest energy source, but it has the capacity to become the first. Photosynthesis can store up to 5-8 times more energy in biomass annually than the actual world energy consumption (Prins et al., 2005). The basic reaction of photosynthesis is as follow: carbon dioxide and water are converted to glucose and oxygen, an endothermic process for which the energy is supplied by photons. Examples of biomass are residues from agriculture or from the forest industry such as branches, straw, stalks, saw dust, etc. An important example of residual agricultural biomass is related to the ethanol production from sugar cane in Brazil which produces 280 kg of residual bagasse at 50% of dry solids. Lignocellulosic materials can be collected and recovered because this material has some energetic content (Ballerini and Alazard-Toux, 2006), however, leaving a part of this material on place is imperative since it keeps the soil fertile.

It is important for governments and citizens to realize that hydrocarbon-based waste material is another source of energy that should be taken advantage on. Waste can be solid or liquid form. It can be land filled, incinerated or converted. Municipal solid waste used electrical transmission poles and railroad ties treated with creosote, sludge from wastewater treatment and pulp and paper industries, wood from construction and demolition operations which contains paints and resins, etc., are all materials that contain carbon that can be valorized in bio-refineries such as the one Enerkem is constructing (2011) in Edmonton, AB.

Assessment of residual biomass or Municipal Solid Waste (MSW) as feedstock to produce bio-ethanol requires a basic understanding of feedstock composition and of the specific properties that dictates its performance as feed in the gasifier. The most important are: moisture content, ash content, volatile matter content, elemental composition and heating value.

The moisture content of biomass is the quantity of water in the material, expressed as percentage of material weight. This weight can be referred to on a wet basis or on a dry basis. If the moisture content is determined on a "wet" basis, the water's weight is expressed as a percentage of the sum of the weight of the water, ash, and dry- ash free matter. It is sometimes necessary to dry the feedstock to a certain level in order to maximize the gasification reaction. Indeed, more moisture is transferred by a higher consumption of oxygen in order to keep the ideal temperature in the gasifier. Temperature of gasification is crucial on the process efficiency. An optimum exists with just the right amount of oxygen needed to perform completely the gasification reaction. This represents a temperature of about 660°C for biomass with 20 % of moisture and about 695°C for biomass with 10 % of moisture (Prins et al., 2005). If more oxygen is added, formation of carbon dioxide will increase and gasification efficiency will drop; the heating value of the synthesis gas will thus decrease (van der Drift et al., 2001). However, not enough oxygen will promote reduction of carbon leading to an increase of methane formation.

The inorganic component (ash content) can be expressed the same way as the moisture content. In general, the ash content is expressed on a dry basis. Both total ash content and chemical composition are both important in regards of the gasification process. The composition of the ash affects its behaviour under high temperatures of combustion and gasification. For example, melted ash may cause problems in both combustion and gasification reactors. These problems may vary from clogged ash-removal caused by slagging ash to severe operating issue in fluidized bed systems. Measurement of ash melting point is thus crucial.

Volatile matter refers to the part of biomass that is released when the biomass is heated beyond its dehydration temperatures. During this heating process the biomass decomposes into volatile gases and solid char. Biomass typically has a high volatile matter content (up to 80%), whereas coal has a low volatile matter content (<20%).

Elemental composition of the ash-free organic component of residual biomass is relatively uniform. The major components are carbon oxygen and hydrogen. Most biomass also contains a small proportion of nitrogen and sulfur. Table 1 presents the elementary composition of biomass as derived from ultimate analyses.

Table 1: Elementary composition of residual biomass

Element	WT% (Dry Basis)
CARBON	44-51
HYDROGEN	5.5-6.7
OXYGEN	41-50
NITROGEN	0.12-0.6
SULFUR	0-0.2

The heating value of a fuel is an indication of the energy chemically bound in the fuel with reference to a standardized environment. The standardization involves the temperature, state of water, and the combustion products. The calorific value is presented as the higher heating value and the lower heating value. The higher heating value represents the heat release per unit of mass when the material (at 25°C) is completely oxidized to carbon dioxide and water and then returned to 25°C. A calorimetric bomb is the standard instrument used to measure this value. The lower heating value is not taking into account the energy supplied by the condensation of water (latent heat of vaporization of water at 25°C which is 2440 kJ/kg). The water includes moisture from the feedstock and the product from the reaction between oxygen and hydrogen comprised in the raw material (Borman and Ragland, 1998). Basic and complementary information about biomass intended for gasification can be obtained via the proximate and ultimate analyses.

Proximate analysis measures moisture content, volatile matter, fixed carbon, ash content and calorific value.

Ultimate analysis provides information about elementary composition of the biomass in weight percentage of carbon, hydrogen, oxygen, sulphur and nitrogen. The carbon to hydrogen ratio in the feedstock has a direct impact on the syngas, more particularly on the ratio of H_2/CO (Higman and van der Burgt, 2008).

Table 2: Properties of biomass feedstocks

ANALYSIS		SAWDUST	BAGASSE	SWITCHGRASS	STRAW	BITUMINOUS COAL	DRY SEWAGE (MSW)
Utlimate Analysis (% W/W) Dry Basis	C	50	48	43	43,5		20.5
	H	6,3	6.0	5.6	4.2		3.2
	O	43	42	46	40.3	6	17.5
	N	0,8	–	0.5	0.6	1.2	2.3
	S	0,03	–	0.1	0.2	5.1	0.6
Proximate Analysis (% W/W) Dry Basis	Moisture	7.8	1	8.4	7.6	8.7	4.7
	Volatiles	74	80	73	68.8	36.1	41.6
	Fixed carbon	25.5	15	13.5	13.5	42	2.3
	Ash	0.03	4	4.5	10.1	21.9	56
	HHV (Mj/Kg)	19.3	17	15.4	17	27	8

Table 2 (depicted below) presents a comparison of different feedstock properties. Moisture content was provided after drying of feedstock and compositions are approximated (Ciferno and Marono, 2002).

Note that higher calorific value formulas, refering to equation (Eq. 1), were developed from the composition of the feedstock. Dulong formula is one of them (Higman and van der Burgt, 2008). However, numerous correlations were developed and reported in literature. For instance, experiments on more than 200 species were performed at the Indian Institute of Technology (Bombay) and the following equation was derived from empirical data.

$$\text{IJ / kg}) = 0.3491C + 1.1783H - 0.1034O - 0.0211A + 0.1005S - (\qquad (1)$$

Where C is the weight fraction of carbon, H of hydrogen, O of oxygen, A of ash, S of sulfur and N of nitrogen. This correlation offers an average absolute error of 1.45 % (Channiwala and Parikh, 2002).

There are other feedstock properties that are critical to determine in order to treat the material accordingly: size, distribution and shapes of the material, porous structure and bulk density.

Gasification and incineration are both design for transforming hydrocarbon-based hazardous material to more simple by-products. Nonetheless, these two technologies are completely different in their operation and finalities. The goal of the combustion is to proceed to a complete oxidation of carbon in order to produce carbon dioxide and water. Thereby an excess of oxidizing agent is used. Sulfur and nitrogen comprised in the feedstock are oxidized to SO_x and NO_x. Whilst only gasification can lead to synthesis, both processes can lead to production of electricity. However, with the goal of generating added-value product, gasification has the possibility to convert the carbon of waste to biofuels instead of releasing the carbon as CO2. Gasification is a partial oxidation process with just enough oxygen to produce the heat to yield reduced carbon and hydrogen (Wetherold et al., 2000).

Syngas Production and Purification

Gasification produces a synthetic gas or syngas mainly composed of hydrogen, methane and carbon monoxide (Marie-Rose et al., 2011, Villano et al., 2010). Gasification started in the late 1800's and first applications were for town gas for street lights illumination and cooking purposes. Other applications were heating, production of raw material for chemical industry and also power generation. When the Second

World War was raging and oil was limited, wood gasification was used for transportation and heat and power generation. The energy crisis of the 70's (1973 and 1979 oil shocks) also provided motivation for improving gasification technologies (Higman and van der Burgt, 2008).

The basics behind gasification are to react the feedstock containing carbon with steam and oxygen (could be air, enriched air or pure oxygen). The following chemical reactions are predominant during gasification (Higman and van der Burgt, 2008)

Thermal decomposition (i.e. pyrolysis), which covers dehydration as well as cracking reactions leading to gases, intermediate vapours and carbon structures known as "char".

Partial oxidation of the "char"which forms CO and CO2 generating heat for the otherwise endothermic reactions.

The partial oxidation

$$C+\frac{1}{2}O_2=CO \quad -111\,MJ/kmol \tag{2}$$

$$C+\frac{1}{2}O_2=CO_2 \quad -283\,MJ/kmol \tag{3}$$

$$H_2+\frac{1}{2}O_2=H_2O \quad -242\,MJ/kmol \tag{4}$$

$$C+H_2O=CO+H_2 \quad 131\,MJ/kmol \tag{5}$$

$$CH_4+H_2O=CO+3H_2 \quad 206\,MJ/kmol \tag{6}$$

$$CxHy+xH_2O=x\,CO+(\frac{1}{2}y+x)H_2 \tag{7}$$

$$CxHy + xCO_2 = 2x\,CO + (\frac{1}{2}y + x)H_2 \tag{8}$$

$$CxHy + xCO_2 = 2x\,CO + (\frac{1}{2}y + x)H_2 \tag{9}$$

$$C + 2H_2 = CH_4 \qquad -75\,MJ / kmol \tag{10}$$

$$CO_2 + H_2 0 = CO_2 + H_2 \qquad -41\,MJ / kmol \tag{11}$$

Steam is hence chemically involved in the gasification, but it also served as moderating agent to limit the temperature. Carbon dioxide could also be used to this purpose. Other elements in feedstocks such as sulphur, nitrogen and chlorine will lead to different molecular species contaminating the syngas such as H_2S, COS, $NH_{3,}$ HCN and HCl.

Different types and configurations of reactors can be used to produce syngas: fixed bed, entrained flow gasifier and bubbling fluidized bed. Fixed bed gasifier (essentially slowly moving-beds) can be updraft or downdraft.

The Updraft fixed bed gasifier is probably the oldest technology. Biomass or coal are fed from the top and move slowly by gravity. A grate supports the material while oxidizing the gas in its way upwards. The synthesis gas is withdrawn at the top of the reactor and ash from the bottom. There is a combustion zone in the bottom where char combusts in contact of oxygen to form CO_2. Char reacts with carbon dioxide to form two moles of carbon monoxide (Eq. 9). A zone of pyrolysis take place a little higher in the fixed bed because all oxygen has been consumed at this level. Finally, hot synthesis gas dries the feedstock at its entrance (Ciferno and Marono, 2002, Paes, 2005). This simple proven process has as drawback the production of a tar-rich syngas. Downdraft fixed bed gasifier employs co-current flow of oxidant and feedstock. This configuration has low tar formation as an outcome. Nevertheless, a part of the carbon remains unconverted (Ciferno and Marono, 2002).

Entrained flow gasifier utilizes a blast of oxidant in a co-current arrangement with the feed grounded to ensure carry-over. A high temperature is necessary because of the short residency time, counted in seconds. More oxygen is thus required to reach higher temperature (Higman and van der Burgt, 2008). With this type of gasifier, the particles entrained are passed to a cyclone and a riser to bring this material again through the bottom of the gasifier.

Bubbling fluidized bed gasifer is designed to keep the particles in suspension. This technology is based on a configuration where feedstock is added continuously to a bed of alumina, olivine (iron and magnesium orthosilicate) and/or other material serving as a heat carrier. This type of gasifier has high heat/mass transfer and conversion and is able to treat a large range of particles sizes and heterogeneous feedstock. Fluidized bed gasifier has the following advantages: flexibility and easiness for control and maintenance. Figure 1 illustrates an overview of the process developed by Enerkem Inc based on a bubbling fluidized reactor.

Feeding is assured by conveyor, lock hoppers, rotary valves and cooled feeding screw. Oxygen and steam are added through the bed in order to fluidize the medium. Low severity conditions are used in the bed, i.e. pressure lower than 4 atm and temperature lower than 750°C (Chornet et al., 2010). The reactor itself is made from carbon steel with insulation and refractory.

Ash is collected at the cyclones and larger inert material exits at the bottom of the gasifier. Ash can be composed of silica, alumina, potassium, phosphorus, sodium, magnesium, ferric oxide, and smaller amounts of titanium oxide and sulfur compounds (Higman and van der Burgt, 2008).

CaO/MgO is also introduced in the gasifier for neutralising a part of HCl, H_2S and COS formed. Literature reports that the use of calcined dolomite and NiO-loaded calcined dolomite helps in the reduction of tar and char formation (Corujo et al., 2010). This dolomite contained 24,6 % of Ca, 19,7 % of Mg and 0,035 % of Fe on a weight basis. Calcination of dolomite produces magnesium and calcium oxides. Corujo et al. have stated that the use of NiO-loaded calcined dolomite catalysts increased the total product gas volume by 30% and decreased the rate of char and tar formation leading to a higher gas energy yield. In fact, bed materials are selected for two criteria: attrition resistance and possible catalytic activity in hydrocarbon and tar reforming. Olivine was compared to silica sand, dolomite, ash, magnetite and iron ore. With the use of Olivine, higher conversion of toluene was observed during steam gasification at 900°C (Rauch et al., 2004).

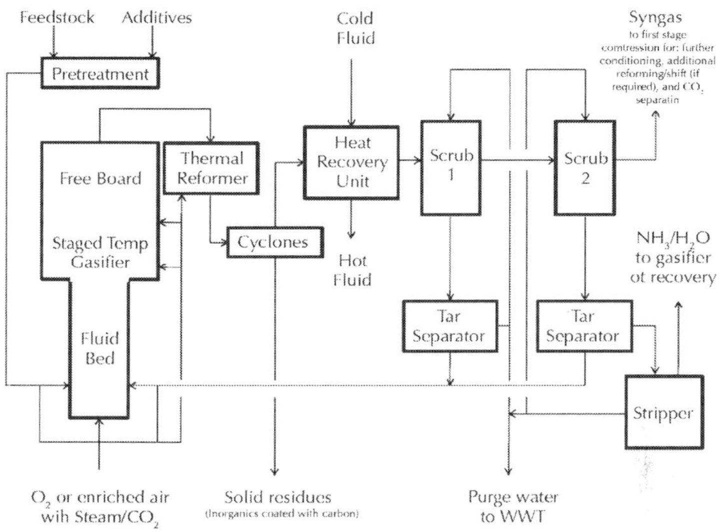

Figure 1: Schematic diagram of Enerkem's process.

Comparison of Ni/olivine catalyst with olivine alone demonstrated a higher activity, a higher selectivity to hydrogen and carbon monoxide and a lower carbon deposition. These observations were explained by an enhanced steam adsorption on magnesium oxide facilitating surface carbon gasification rates and by nickel dilution effect, associated with Fe-Ni alloys, ensuring stable catalyst activity (Swierczynski et al., 2007)

Leaving the freeboard of the gasifier, the syngas is composed of: H_2, CO, CO_2, H_2O N_2, hydrocarbons, tar, fines, char and contaminants. Tar is a complex blend of condensable hydrocarbons, which includes single to five-rings aromatic compounds along with oxygen-containing hydrocarbons and complex PAH (Yan et al., 2006). The freeboard is sized appropriately for disengagement of alumina and/or other solid fluidized bed materials.

A reforming zone is used for converting hydrocarbons, tar and char to obtain more syngas (see Eq. 8,Eq. 9, Eq. 10 and Eq. 11). Since thermal decomposition of hydrocarbons requires activation and that reforming reactions are endothermic, heating the gas, by direct or indirect manner, is essential. For example, if oxygen is injected directly, it must be added carefully to avoid lowering the product gas heating value and cold gas efficiency. However, with just the right amount of the oxidizing gas, selective oxidation of methane, ethylene and propylene will occur (Villano et al., 2010). Target temperatures are in the 750°C to 1200°C range. Table 3 depicts general compositions of syngas produced out of some previously mentioned biomass.

Table 3: Solid, liquid and gas ratio as well as gas composition following gasification of common residual feedstocks

BIOMASS	MSW	MSW	SLUDGES*	RES. FOREST/ AGRICULTURE	TYPICAL BIOMASS
CONDITIONS	700 °C, FIXED BED GASIFIER, DOLOMITE (CATALYST),	900 °C, FIXED BED GASIFIER, DOLOMITE (CATALYST), STEAM (OXIDIZING AGENT)	829 °C FLUIDIZED BED	900 °C, FIXED BED GASIFIER, DOLOMITE (CATALYST), STEAM (OXIDIZING AGENT)	O_2 / STEAM / CO_2 LOW SEVERITY GASIFICATION + THERMAL CRACKING AND REFORMING AT HIGHER SEVERITY
CHAR (%W/W)	19.15	12.65	N.D.	N.D.	N.D.
TAR (%W/W)	12.94	2.62	N.D.	N.D.	N.D.
GAS (%W/W)	94.52	145.23	N.D.	N.D.	N.D.
GAS COMPOSITION	%mol	%mol	%mol	%mol	%mol
H_2	16.92	36.98	17.23	46.2	23-27
CO	20.33	27.37	26.17	33.2	21-23
CO_2	35.28	20.78	45.64	16.1	38-44
CH_4	21.44	9.94	7.96	4.4	6-8
C_2	6.03	4.93	3.00	0.1	–
C_2H_4	n.d.	n.d.	n.d.	n.d.	Traces
C2-C5	N.D.	N.D.	N.D.	N.D.	0.2-0.5
C5-C10	N.D.	N.D.	N.D.	N.D.	TRACES

Increasing temperature cause the decomposition of char and tar by thermal cracking and steam reforming reactions. Moreover, this has a direct impact on the production and composition of the gas as seen from table 3 (He et al., 2009)

Cyclones are used for entrained ash removal. The latter having typically a diameter greater than 10 microns. Efficiency of this cyclone system is from 90 to 95% and mostly recuperates char. Following the cyclone and recuperation of ash, the next unit operation is a heat exchanger. The latter cools the crude synthesis gas and allocates heat recovery for other application on the process using thermal oil or liquid water which turns to steam.

Note that the reforming step mentioned above could be by-passed directly to the cyclones if the feedstock contains inorganic materials that form eutectics with low melting point Indeed, high temperature in this situation would have as a consequence scaling and fouling of the walls of the system.

Quenching is required for further cooling and removal of condensable materials, tars and fines. In the following step, a wet venturi scrubber with alkaline water is used for neutralisation of acid gases as H_2S and HCl. A coalescer/demister at the exit removes fines particles and mist.

A second scrubbing step involving neutral or slightly acidic water is used in order to capture ammonia, trace tars, residual fines and impurities such as chlorine and metals. After this cooling step, the cooled purified syngas is around 30 °C.

Water is recirculating after proper handling. The water loop includes knockout drums and separation of tars and fines particles with air flotation, decanters and/or centrifuges. The skimmed phase recovered (overflow), enclosing tars and fines particles, is emulsified and recycled to the gasifier. The underflow, containing heavier organics and particles, is also re-injected to the gasifier. A part of the recirculated water is withdrawn from the system to maintain the balance of water and contaminants. This purge is subjected to water treatment in the objective of meeting below the required environmental standards. Note that ammonia is separated from the water by steam stripping above 100°C and at about 1 to 3 atm. The ammonia is sent back to the gasifier producing hydrogen and nitrogen.

An additional reforming step is then possible for converting light hydrocarbons to more CO and $H_2$and final adjustment of the molar ratio of H_2/CO.

The syngas is then passed through filters and adsorbers for dehumidification, elimination of traces of chlorine, sulphur and other contaminants such as metal carbonyls. An utlraclean syngas is obtained and compression follows to remove carbon dioxide assisted either by amine-based processes or chilled methanol.

SYNGAS CONVERSION INTO METHANOL

Thermodynamic Consideration

The two major components of synthesis gas, hydrogen and carbon monoxide are the building blocks of what is often known as C1 chemistry. Conversion of *syngas* to liquid fuels as well as conversion rates is directly related to the composition of the catalyst. Syngas can be efficiently converted to different products as alcohols and aldehyde (Figure 2).

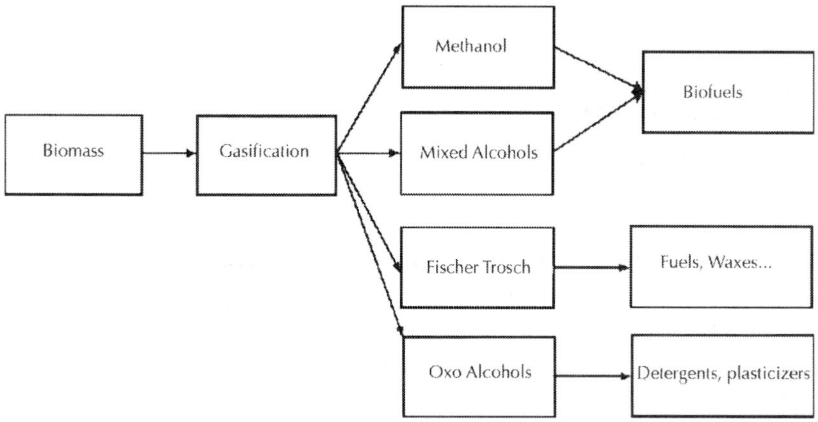

Figure 2: Examples of application for syngas produced from biomass (Higman and van der Burgt, 2008).

Although many routes are available, the most promising route at industrial level is the production of methanol since the synthesis yields are the highest. The conversion to synthetic liquids as methanol is strongly influenced by thermodynamic factors (Rostrup-Nielsen, 2000).

$$CO + 2H_2 = CH_3OH \qquad -90.64\,MJ/kmol \tag{12}$$

$$CO_2 + 3H_2 = CH_3OH + H_2O \qquad -49.67\,MJ/kmol \tag{13}$$

$$CO + H_2O = CO_2 + H_2 \qquad -41.47\ KJ/mol \tag{14}$$

For methanol synthesis, a stoichiometric ratio defined as $(H_2\text{-}CO_2)/(CO+CO_2)$, of about 2 is preferred, which implies that there should be just the stoichiometric amount of hydrogen needed for methanol synthesis. For kinetic reasons and in order to control by-products formation, a value slightly above 2 is normally preferred (Dybkjr and Christensen, 2001). Moreover, methanol synthesis is subjected to a thermodynamic equilibrium that limits the process to low conversion per pass and therefore implies a large recycle of unconverted gas. The reaction is strongly exothermic and consequently requires significant cooling duty.

Different phenomena exist at high and low pressure conditions. As example, when the pressure is relatively low, increasing temperature, CO conversion is not monotonic, and the trend is that of an increase followed by a decrease with the maximum conversion appearing near 250°C. This phenomenon is in agreement with many works in the literature (Li and Inui, 1996, Liaw and Chen, 2001, Wang et al., 2002). As the reaction temperature increases, the reaction rate gets higher and leads to the increase of CO conversion. However, methanol synthesis is an exothermic reaction and low temperature is more beneficial considering equilibrium. The conversion does not continue to increase due to the thermodynamic limitation and a decrease trend will even appear. When the system pressure is relative high, because of the relatively low CO conversion, the system is far from the thermodynamic equilibrium and under the control of reaction kinetics. In this case, CO conversion increases monotonically with an increase in temperature. From a mechanistic point of view for the methanol synthesis, two main reactions need to be in line with real world situation: CO hydrogenation and CO_2 hydrogenation.

CO HYDROGENATION AS PRINCIPAL REACTION FOR METHANOL SYNTHESIS

In this case, the main reaction leading to methanol is the hydrogenation of CO.

$$CO + 2H_2 = CH_3OH \qquad -90.64 \text{ KJ/mol} \qquad (15)$$

$$CO_2 + H_2 = CO + H_2O \qquad 41.14 \text{ KJ/mol} \qquad (16)$$

According to equation (Eq. 15) methanol is predominantly synthesised via the direct hydrogenation of CO. The second reaction is the reverse of Water Gas Shift (RWGS). Experimental data involving typical syngas mixtures that contain 3 to 9% CO_2 show a decrease of its concentration in the reactor effluent stream (Sunggyu, 2007). It should be noted that the first reaction (methanol synthesis) is exothermic, whereas the second (RWGS) is endothermic. According to this depletion of carbon dioxide in the RWGS reaction, produce more reactant (CO) which overall boost the synthesis of methanol. Until 90s the role of CO_2 in the methanol synthesis was not clear. Deficiency of CO_2 in the feed composition can be extremely detrimental to the overall synthesis, very rapidly deactivating the catalysts and immediately lowering methanol productivity in the process. Typically, 2 to 4% of CO_2 is present in the syngas mixture for the vapor-phase synthesis of methanol, whereas this value is somewhat higher, with 4 to 9%, for liquid-phase synthesis (Cybulski, 1994).

$$CO_2 + 3H_2 = CH_3OH + H_2O \qquad -49.67 \text{ KJ/mol} \qquad (17)$$

$$CO + H_2O = CO_2 + H_2 \qquad -41.14 \text{ KJ/mol} \qquad (18)$$

In this case, the main reaction leading to methanol is hydrogenation of CO_2.

It should be noted that according to this reaction, the synthesis of methanol proceeds predominantly by direct hydrogenation of CO_2 and not CO. It should also be noted that the WGS reaction proceeds in the forward direction, consuming CO to produce the main reactants: CO_2 and H_2, thus boosting the eventual methanol productivity. A number of authors (Cybulski, 1994, Lee et al., 1989) have tried a variety of reaction to elucidate the true reaction pathways or mechanistic pathways, including isotope labelling studies and kinetic studies involving complete absence of one of the syngas components.

Catalysts Used

The first high-temperature, high-pressure ever used methanol synthesis catalysts were $ZnO/Cr_2O_3$and were operated at 350°C and 250-350 bar. Catalyst compositions contained 20-75 atom% Zn and these catalysts demonstrated high activity and selectivity for methanol synthesis and proved robust enough resist sulphur poisoning which is inherent when converting syngas from coal gasification. Over the years, as gas purification technologies improved, interest in the easily poisoned Cu catalysts for methanol synthesis was renewed. In 1966, ICI introduced a new, more active $Cu/ZnO/Al_2O_3$catalyst was the first of a new generation of methanol production using lower temperature (220-275°C) and lower pressure (50-100 bar) than the established ZnO/ Cr_2O_3 catalysts. The last high temperature methanol synthesis plant was closed in the mid-1980s (Fiedler et al., 2003) and at the present, low temperature, low pressure processes based on Cu catalyst are used for all commercial production of methanol from syngas. The synthesis process has been optimised to the point that the modern methanol plants yield 1 kg of methanol/liter of catalyst/hr with >99.5% selectivity for methanol. Commercial methanol synthesis catalyst has lifetimes on the order of 3-5 years under normal operating conditions.

The Cu crystallites in methanol synthesis catalysts have been identified as the active catalytic sites although the actual state (oxide, metallic...) of the active Cu site is still being debated. The most active catalysts all have high Cu content, optimum about 60 wt% Cu on the catalyst that is limited by the need to have enough refractory oxide to prevent sintering of the Cu crystallites. Hindering agglomeration is why ZnO creates a high Cu metal surface area. ZnO also interacts with Al_2O_3 to form a spinel that provides a robust catalyst support. Acidic materials like alumina, are known to catalyse methanol dehydration reactions to produce DME. By interacting with the Al_2O_3 support material, the ZnO effectively improves methanol selectivity by reducing the potential for DME formation. Catalysts are typically prepared by the co-precipitation of metal salts with a variety of precipitation agents. It is important to avoid contaminating methanol catalysts with metals that have FT (Fischer Tropsch) activity (Fe or Ni) during the synthesis. Incorporation of alkali metal in the catalyst formulation should also be avoided because they catalyse the increase of higher alcohols production. Table 4 shows catalyst formulation from several commercial manufacturers.

Additional catalyst formulations have been presented in the literature with the purpose of improving per-pass methanol yields (Klier, 1982). The addition of Cs to Cu/ZnO mixtures has shown improved methanol synthesis yields. This only holds true for the heavier alkali metals, as the addition of K to methanol synthesis catalysts tends to enhance higher alcohols yields. The Cu/ThO_2 intermetallic catalysts have also been investigated for methanol synthesis (Klier, 1982). These catalysts have demonstrated high activity for forming methanol from CO_2-free syngas. Cu/Zr catalysts have proven active for methanol synthesis in CO-free syngas at 5 atm and 160-300°C (Herman, 1991). Supported Pd catalysts have also demonstrated methanol synthesis activity in CO_2-free syngas at 5-110 atm and 260-350°C (Spath and Dayton, 2003).

Table 4: Commercial Methanol synthesis catalyst formulation (Spath and Dayton, 2003)

Manufacter	**Cu** (wt%)	**Zn** (wt%)	**Al** (wt%)	**Other**
IFP	45 – 70	15 – 35	4 – 20	Zr – 2-18
ICI	20 – 35	15 – 50	4 – 20	Mg
BASF	38.5	18.6	12.9	Rare Earth oxide – 5
Shell	71	24		
Sud Chemie	65	22	12	
Dupont	50	19	31	
Haldor Topsoe	"/>55	21 - 25	8 - 10	

Methanol Synthesis Reactors

Today, the majority of the methanol is synthesised from syngas produced by steam reforming of natural gas (SMR). The synthesis can be done either with heat provided by a furnace where the tubular reactor is located, or by auto-thermal reforming (ATR) combined with steam reforming. Once the natural gas is reformed the resulting synthesis gas can be shift adjusted for its H2/CO ratio and the CO2 decreased to a few percentages as previously specified. The syngas is then fed to a reactor vessel in the presence of a catalyst to produce methanol and water vapor. This crude methanol, which usually contains up to 18% water, plus ethanol and higher alcohols is fed to a distillation plant that consists of a unit that removes the volatiles and a unit that removes the water and higher alcohols. The unreacted syngas is recirculated back to the methanol converter resulting in an overall conversion efficiency of 99%. A generic methanol synthesis process flow diagram (PFD) is shown in the Figure 3.

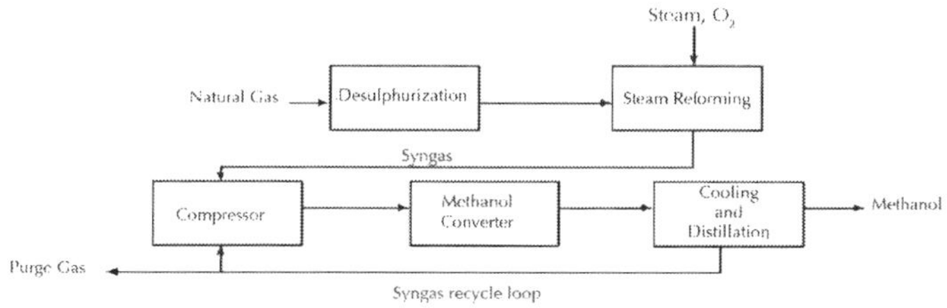

Figure 3: Simplified Methanol Synthesis PFD (Spath and Dayton, 2003).

As is the case with Fisher Tropsch synthesis, one of the challenges associated with commercial methanol synthesis is removing the large excess heat of reaction. Methanol synthesis increases at higher temperatures so does the chance for competing side reactions. Controlling and dissipating the heat of reaction and overcoming the equilibrium constraint to maximise the per-pass conversion efficiency are the two main process features that are considered when designing the methanol synthesis reactor, commonly referred to as a methanol converter. Numerous methanol converter designs have been commercialised over the years and these can be roughly separated into two categories: adiabatic or isothermal reactors.

- Adiabatic reactors often imply multiple catalysts beds separated by gas cooling devices, either direct heat exchange or injection of cooled, fresh or recycled syngas.
- The isothermal reactors are designed to continuously remove the heat of the reaction so they operate essentially like a heat exchanger.

HaldorTopsoe low-pressure methanol synthesis process. This process is designed to produce methanol from natural gas or associated gas feedstocks, utilizing a two-step reforming process to generate a syngas mixture feed for the methanol synthesis (Sunggyu, 2007). Associated gas is natural gas produced with crude oil from the same reservoir. It is claimed that the total investment for this process is lower than with the conventional flow scheme based on straight steam reforming of natural gas by approximately 10%, even after considering an oxygen plant. As shown in figure 4, the two stage reforming is conducted by primary reforming in which a preheated mixture of natural gas and steam is reacted, followed by a secondary reforming which further converts the exit gas from the primary reformer with the aid of oxygen that is feed separately. The amount of oxygen required as well as the balance of conversion between the primary and the secondary reformers need to be properly adjusted so that a balanced syngas (in a stoichiometric ratio (2:1) of H_2/CO) is obtained with a low inert content.

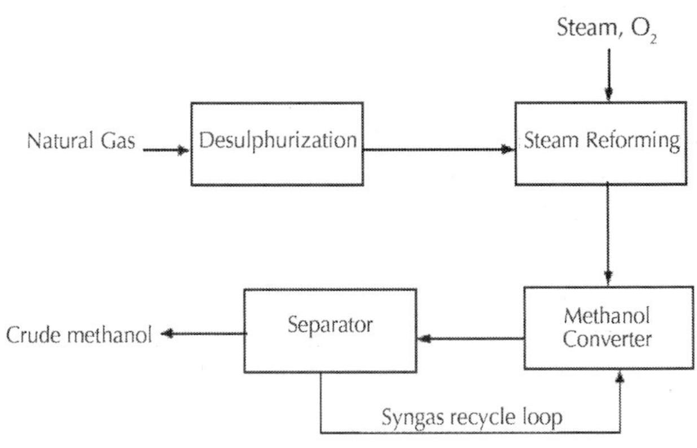

Figure 4: Haldor Topsoe methanol synthesis process (Sunggyu, 2007).

Liquid-Phase methanol process. The liquid phase methanol process was originally developed by Chem. Systems Inc in 1975 (Cybulski, 1994). The R&D of this process was sponsored by the U.S. Department of Energy and Electric Power Research Institute. Commercialised by Air Products and Chemicals Inc and Eastman Chemical Co. in the 1990s, the process is based on the low-pressure methanol synthesis using coal as the source of syngas. Recently, in Québec (Canada) Enerkem Inc. has developed a liquid phase methanol process using syngas produced from biomass. The chemical reaction is carried out in a slurry reactor using a $Cu/ZnO/Al_2O_3$ catalyst at temperature ranging from 230 to 260 °C and 50 to 100 atm. The commercial reactor used a liquid entrained reactor in which fine grains of catalyst are slurried in an inert high-boiling oil typically white mineral oil. Pressurized gaseous reactants are dissolved in the oil and the dissolved molecular species are reacted on the catalytic surfaces of the grains present in a slurry. The figure 5 shows the schematic of liquid phase methanol process of Enerkem Inc.

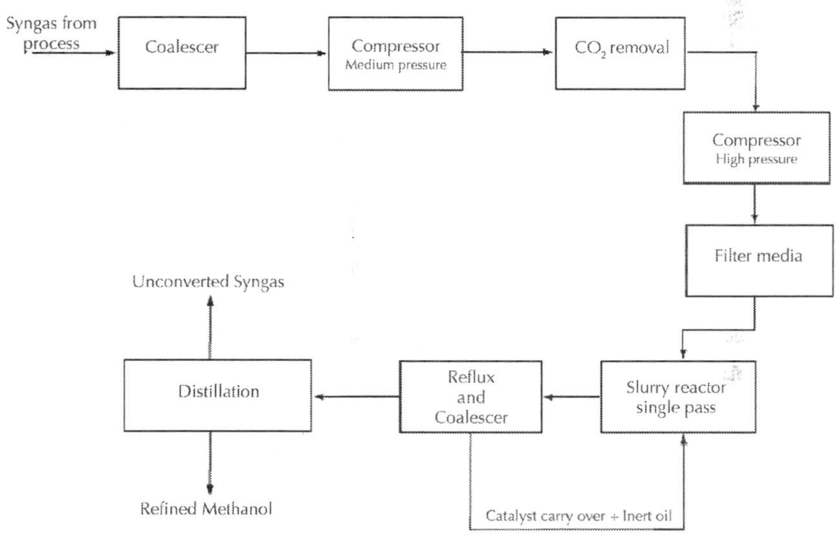

Figure 5: Enerkem Inc. liquid phase methanol process.

ETHANOL SYNTHESIS

Ethanol can be readily produced by fermentation of simple sugars that are hydrolyzed form starch crop. Feedstocks for such fermentation include corn, barley, potato, rice and wheat (Cybulski, 1994). Sugar ethanol can be called *grain ethanol*, whereas ethanol produced from cellulose biomass is called*cellulosic or second generation ethanol*. Both grain ethanol and cellulosic are produced via biochemical processes, whereas *chemical ethanol* is synthesised by chemical synthesis routes that do not involve fermentation. In fact, thermo-catalysis process route can be a good alternative to the bioprocess route. In a recently reported process (Chornet et al., 2009) the ethanol is produced by a two steps process: first methanol carbonylation to produce methyl acetate and second hydrogenolysis of the latter as illustrated in the figure 6.

Methanol Carbonylation: Methyl Acetate Synthesis

Methyl acetate is produced industrially by methanol carbonylation in the acetic acid production processes. Methanol carbonylation reaction is a well known reaction since Monsanto in 1970 performed it in liquid phase using a rhodium-base catalytic system. which was later acquired and optimized by Celanese. Such system had been preceded by a cobalt-based system developed by BASF, which suffered from significantly lower selectivity and the necessity for much harsher conditions of temperature and pressure. Although the rhodium catalysed system has much better activity and selectivity, the R&D has continued for new catalysts which improve efficiency even further. The strategies employed have involved either modifications to the rhodium-based system or insertion of another metal, eventually Iridium (Haynes, 2006). In the homogenous system, acetic acid was used as solvent, so esterification leads to substantial conversion into methyl acetate. Methyl acetate formation is facilitated by the methyl iodide used in carbonylation as a mediator. Heterogenous systems for methanol carbonylation have been suggested for several years. Research has been primarily focus on two possible catalysts for this reaction: 1) rhodium supported polymers (Acetica process as an example) or 2) zeolites and a variety of metals supported on activated carbon (Merenov and Abraham, 1998). The choice of the support seems to play an important role in the activity of the catalyst (Yashima et al., 1979). Another way to produce methyl acetate is by an esterification reaction. Eastman Chemical Company, produced methyl acetate by the esterification of methanol with acetic acid in a reactive distillation column.

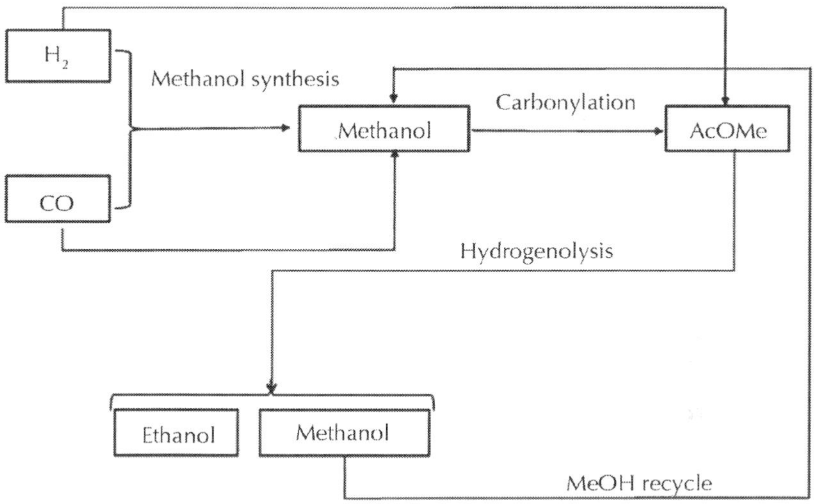

Figure 6: Enerkem Inc. ethanol pathway production.

This company claimed the first commercial plant of reactive distillation (Zoeller, 2009). As illustrated by figure 7, in the reactive distillation there are three sections in the column, water/methanol stripping zone in the bottom, a reaction zone in the center, and methyl acetate enriching at the top. Descriptively, methanol (b.p. 65°C) is added at the bottom of the reaction zone and acetic acid (b.p. 118°C) and acid catalyst are added at the top of the reaction zone.

As methanol ascends up the column it encounters a consistently richer acetic acid concentration, which drives the equilibrium toward the products and prevents methanol from leaving column. Likewise, as acetic acid descends down the column, it also encounters a continuously enriched stream of methanol, which also pushes the equilibrium toward products and prevents acetic acid reaching the bottoms of the column. However, the acetic acid also serves as extractant for water, breaking the various azeotropes. A short section is left at the base of the distillation column to strip all the methanol and a short section is added at the top of the column to enrich methyl acetate and prevent acetic acid from being entrained as the product is distilled overhead in the column. By properly balancing the addition rates, the reaction provides methyl acetate in 100% yield based on both acetic acid and methanol.

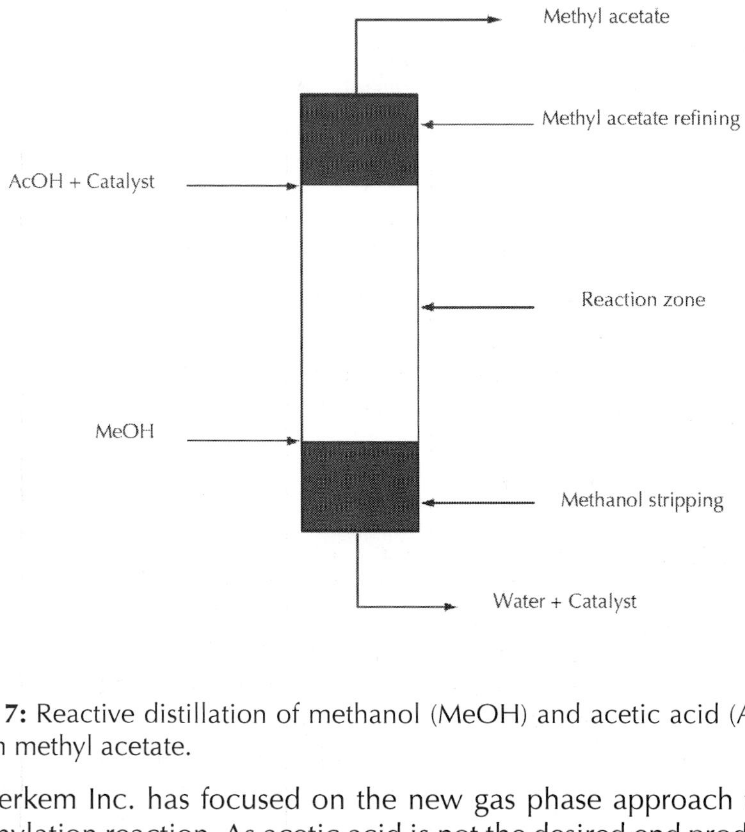

Figure 7: Reactive distillation of methanol (MeOH) and acetic acid (AcOH) to form methyl acetate.

Enerkem Inc. has focused on the new gas phase approach to the carbonylation reaction. As acetic acid is not the desired end product of the carbonylation, the reaction conditions have been tailored to favour the methyl acetate directly according to the equation:

$$2CH_3OH + CO = CH_3COOCH_3 + H_2O \quad -138.08 \text{ KJ/mol} \qquad (19)$$

This is advantageous for Enerkem Inc. since it reduces by one step their process leading to ethanol. Forcing the reaction towards methyl acetate also facilitates a strategy based on gas phase reactor. In the process, the methanol carbonylation is carried out maintaining methanol in the vapor phase using a fixed bed packed a rhodium-based catalyst. The methanol is vaporized under pressure and mixed with the CO-rich gas prior to flowing through the reactor. The methanol to CO molar ratio is comprised between 1 and 4. Methyl iodide (co-catalyst and mediator) is added to the system at a suitable mol ratio relative to methanol. The operating conditions are such that the GHSV (based on CO) varies up to 2,000 h^{-1}. At temperature comprised between 170 to 300°C and total pressure from 10 to 50 atm, it is found that the CO is converted at rate near 100% when the methanol:CO ratio is up to 2 (Chornet et al., 2009).

Methyl Acetate Hydrogenation

Esters (i.e. methyl acetate) are ubiquitous in nature and have vast industrial and commercial applications based on different substituent groups. An important class of industrial catalytic process for esters reduction is metal-catalyzed hydrogenolysis for the production of alcohols (e.i. ethanol) (Adkins and Folkers, 1931, Adkins et al., 1933, Turek et al., 1994a, Xu and Xu, 2010) New applications for ester hydrogenolysis are being found in catalytic upgrading and conversion of renewable resources such as lignocellulosic biomass into fuels or fine chemicals (Corma et al., 2007,Huber et al., 2006). Copper-based catalysts are widely used for ester hydrogenolysis (Adkins and Folkers, 1931, Cybulski et al., 2001, Thomas et al., 1992, Turek et al., 1994b, Turek et al., 1994a). They show very good activity and selectivity for alcohols under high temperature and hydrogen pressure (500-700K, 200-300 bar). Precious metal-based alloys (Rh-Sn for example) show promise of lessening the severity of the reaction conditions (lowering hydrogen pressure to below 100 bar), but precious metal are costly and tend to favor hydrocarbon production.

In our approach, the recovered methyl acetate is maintained in liquid phase at 20°C. It is pumped against pressure ranging from 10 to 50 atm, through a heat exchanger that vaporizes it completely at temperature ranging 150 to 425°C. Hydrogen, preheated at the same temperature range is mixed with the methyl acetate vapor at the exit of the heat exchanger. The molar ratio hydrogen to methyl acetate is from 4 to 11. The hot mixture is flown through a catalytic bed where the catalyst CuO/Cr_2O_3 or $CuO/ZnO/Al_2O_3$ catalyst is placed. The CuO is reduced with H_2/N_2 mixtures prior to adding any methyl acetate (Claus et al., 1991). Methyl acetate is converted to ethanol and methanol according to the equation:

$$CH_3COOCH_3 + 2H_2 = CH_3OH + CH_3CH_2OH \quad -27 \text{ KJ/mol} \qquad (20)$$

The GHSV (based on H_2) of the reaction is comprised between 1,000 and 2,000 h^{-1} and the methyl acetate conversion reached 95%, with a slightly higher selectivity towards ethanol.

CONCLUSIONS

Biomass is a renewable energy source whose conversion to biofuels is an option to reduce oil dependency and reduce the carbon dioxide footprint characteristic of fossil fuels.

This chapter has shown the chemical steps needed to convert non-homogeneous biomass into bioethanol via gasification. The syngas produced is catalytically turned into biofuels (methanol and ethanol). Such approach is practices by Enerkem's which uses non-homogeneous residual biomass including urban biomass as feedstock for the gasification. The syngas produced is cleaned prior used for catalytic synthesis of methanol. The reactor used for the methanol synthesis is a three phase reactor based on the Liquid-Phase methanol process. The yield of methanol is about 1 $kg_{MeOH}/kg_{cat}/h$ and the selectivity of the reaction into methanol is about 99%. The refined methanol produced is carbonylated to produce methyl acetate. The synthesis of methyl acetate is carried in a fixed bed reactor and uses a halide component as co-catalyst. The methyl acetate is hydrogenolized to form stoichiometrically one mole of ethanol and one mole of methanol. The methanol produced at the end of the process is recycled into the carbonylation unit.

ACKNOWLEDGEMENTS

The authors would like to acknowledge financial support of the sponsors of the Industrial Chair in Cellulosic Ethanol: the Quebec's government (MRNF) and its industrial partners: Greenfield Ethanol, CRB innovations and Enerkem. Additional support from NSERC's Synergy program is also acknowledged.

REFERENCES

1. J. , L. Rastoin, 2008 Agrocarburant et sécurité alimentaire: l'alliance impossible, Vol.
2. M. J. Prins, K. J. Ptasinski, F. J. J. G. Janssen, 2005 Exergetic optimisation of a production process of Fischer-Tropsch fuels from biomass, 4 375389 , 0378-3820
3. D. Ballerini, N. Alazard-Toux, 2006 Les biocarburants: état des lieux, perspectives et enjeux du développement, Technip, 9782710808695,
4. A. van der Drift, J. , van der Drift, J. van Doorn and J. W. Vermeulen, 2001 Ten residual biomass fuels for circulating fluidized-bed gasification, 1 4556 , 0961-9534
5. G. L. Borman, K. W. Ragland, 1998 Combustion engineering, McGraw-Hill, 9780070065673,
6. C. Higman, M. van der Burgt, 2008 Gasification (2nd Edition), Elsevier, 978-0-7506-8528-3,
7. J. P. Ciferno, J. J. Marano, 2002 Benchmarking Biomass Gasification Technologies for Fuels, Chemicals and Hydrogen Production, report for US DOE.
8. S. A. Channiwala, P. P. Parikh, 2002 A unified correlation for estimating HHV of solid, liquid and gaseous fuels, 8 10511063 , 0016-2361
9. B. Wetherold, D. Orr, D. Maxwell, 2000 A comparison of gasification and incineration Hazardous wastes. US DOE, DCN 99.803931.02
10. S. C. Marie-Rose, J. , M. Lavoie, D. Lynch, E. Chornet, 2011 From biomass-rich residues into fuels and green chemicals via gasification and catalytic synthesis, WIT Transaction on Ecoloy and the Environment, 143 123132 , 1743-3541
11. S. M. Villano, J. Hoffman, H. , H. Carstensen, A. M. Dean, 2010 Selective Oxidation of Ethylene in a Model «Dirty Synthesis

Gas Stream: An Experimental and Kinetic Modeling Study, ACS National meeting, San Francisco

12. T. Paes, 2005 Modeling for control of biomass gasifier, Thesis of Endhoven University of Technology.
13. E. Chornet, B. Valsecchi, G. Drolet, M. Gagnon, B. Nguyen, 2010 Production and conditioning of synthesis gas obtained from biomass. Patent Application 20100051875
14. A. Corujo, L. Yermán, B. Arizaga, M. Brusoni, J. Castiglioni, 2010 Improved yield parameters in catalytic steam gasification of forestry residue; optimizing biomass feed rate and catalyst type, 12 16951702 , 0961-9534
15. R. Rauch, H. Hofbauer, K. Bosch, I. Siefert, C. Aichernig, H. Tremmel, K. Voigtlaender, R. Koch, R. Lehner, 2004 Steam Gasification of Biomass at CHP Plant in Guessing-Status of the Demonstration Plant. 2nd World Conference and Technology Exhibition on Bioamass for Energy, Italy.
16. D. Swierczynski, S. Libs, C. Courson, A. Kiennemann, 2007 Steam reforming of tar from a biomass gasification process over Ni/olivine catalyst using toluene as a model compound, 3-4 ,211222 , 0926-3373
17. C. Yan, W. Yang, J. T. Riley, W. , P. Pan, 2006 A novel biomass air gasification process for producing tar-free higher heating value fuel gas, Elsevier, Amsterdam, PAYS-BAS
18. M. He, B. Xiao, Z. Hu, S. Liu, X. Guo, S. Luo, 2009 Syngas production from catalytic gasification of waste polyethylene: Influence of temperature on gas yield and composition, 3 13421348 , 0360-3199
19. J. R. Rostrup-Nielsen, 2000 New aspects of syngas production and use, 2-4 ,159164 , 0920-5861
20. I. Dybkjr, T. S. Christensen, 2001 Syngas for large scale conversion of natural gas to liquid fuels, Studies in Surface Science and Catalysis, J. J. S. E. Iglesia and T. H. Fleisch,435440 , Elsevier, 0167-2991,
21. J. L. Li, T. Inui, 1996 Enhancement in methanol synthesis activity of a copper/zinc/aluminum oxide catalyst by ultrasonic treatment during the course of the preparation procedure, 1-2 ,8796 , 0926-860X

22. B. J. Liaw, Y. Z. Chen, 2001 Liquid-phase synthesis of methanol from CO2/H2 over ultrafine CuB catalysts, 2 245256 , 0926-860X
23. Y. W. Wang, Z. L. Wang, J. F. Wang, Y. Jin, 2002 Study on the methanol synthesis in slurry reactors, 8 597601 , 10008144 (ISSN)
24. L. Sunggyu, 2007 Methanol Synthesis from Syngas, Handbook of Alternative Fuel Technology, Taylor & Francis Group,
25. A. Cybulski, 1994 Liquid-phase methanol synthesis: Catalysts, mechanism, kinetics, chemical equilibria, vapor-liquid equilibria, and modeling- A review, 4 557615 , 01614940 (ISSN)
26. S. Lee, V. R. Parameswaran, I. Wender, C. J. Kulik, 1989 Roles of carbon dioxide in methanol synthesis, 8 10211057 ,
27. E. Fiedler, G. Grossmann, D. B. Kersebohm, G. Weiss, C. Witte, 2003 Ullmann's Encyclopedia of Industrial Chemistry Release 2003, Methanol, W.-V. V. G. C. KGaA,pp.
28. K. Klier, 1982 Methanol Synthesis, Advances in Catalysis, H. P. D.D. Eley and B. W. Paul,243313 , Academic Press, 0360-0564,
29. R. G. Herman, 1991 Chapter 7 Classical and Non-Classical Routes for Alcohol Synthesis, Studies in Surface Science and Catalysis, L. Guczi,265349 , Elsevier, 0167-2991,
30. P. L. Spath, D. C. Dayton, 2003 Preliminary Screening and Economy Assessment of Synthesis fuels and Cjemicals with Emphasis on the Potential for Biomass-Derived Syngas
31. A. Haynes, 2006 Acetic Acid Synthesis by Catalytic Carbonylation of Methanol, Catalytic Carbonylation Reactions, M. Beller,179205 , Springer Berlin / Heidelberg,
32. A. S. Merenov, M. A. Abraham, 1998 Catalyzing the carbonylation of methanol using a heterogeneous vapor phase catalyst, 4 397404 , 0920-5861
33. T. Yashima, Y. Orikasa, N. Takahashi, N. Hara, 1979 Vapor phase carbonylation of methanol over RhY zeolite, 1 5360 , 0021-9517
34. J. R. Zoeller, 2009 Eastman Chemical Company's «Chemicals from Coal» program: The first quarter century, 3-4 ,118126 , 0920-5861
35. H. Adkins, K. Folkers, 1931 The catalytic hydrogenation of Esters to alcohols, 3 10951097 , 0002-7863

36. E. Chornet, B. Valsecchi, Y. Avilla, B. Nguyen, J. -M, Lavoie, 2009 Production of ethanol to methanol, patent Application, WO20090326080
37. H. Adkins, B. Wojcik, L. W. Covert, 1933 The Catalytic Hydrogenation of Esters to Alcohols. III, 4 16691676 , 0002-7863
38. T. Turek, D. L. Trimm, N. W. Cant, 1994a The Catalytic Hydrogenolysis of Esters to Alcohols, 4 645683 , 0161-4940
39. L. Xu, Y. Xu, 2010 Activation of methyl acetate on Pd(111), 11-12 ,887892 , 0039-6028
40. A. Corma, S. Iborra, A. Velty, 2007 Chemical Routes for the Transformation of Biomass into Chemicals, 6 24112502 , 0009-2665
41. G. W. Huber, S. Iborra, A. Corma, 2006 Synthesis of Transportation Fuels from Biomass:,Äâ Chemistry, Catalysts, and Engineering, 9 40444098 , 0009-2665
42. A. Cybulski, J. Chrzszcz, M. V. Twigg, 2001 Hydrogenation of dimethyl succinate over monolithic catalysts, 1-4 ,241245 , 0920-5861
43. D. J. Thomas, J. T. Wehrli, M. S. Wainwright, D. L. Trimm, N. W. Cant, 1992 Hydrogenolysis of diethyl oxalate over copper-based catalysts, 2 101114 , 0926-860X
44. T. Turek, D. L. Trimm, D. S. Black, N. W. Cant, 1994b Hydrogenolysis of dimethyl succinate on copperbased catalysts, 1-2 ,137150 , 0926-860X
45. P. Claus, M. Lucas, B. Lücke, T. Berndt, P. Birke, 1991 Selective hydrogenolysis of methyl and ethyl acetate in the gas phase on copper and supported Group VIII metal catalysts, 1 118 , 0926-860X

Chapter 5

Scale-Up of a Cold Flow Model of Ficfb Biomass Gasification Process to an Industrial Pilot Plant–Example of Dynamic Similarity

Jernej Mele[1]

[1] Faculty of mechanical engineering/Bosio d.o.o., Slovenia

INTRODUCTION

In this chapter we are introducing the research of particles hydrodynamics in a cold flow model of Fast Internal Circulating Fluidized Bed (FICFB) biomass gasification process and its scale-up to industrial pilot plant. A laboratory unit has been made for the purposes of experimental research. The laboratory unit is three times smaller than the later pilot plant. For a reliable observation of the flow process, similar flow conditions must be created in the laboratory unit and the pilot plant. The results of the laboratory model will be similar to those of the actual device if geometry, flow and Reynolds numbers are the same. Therefore, there is no need to bring a full-scale gasificator into the laboratory and actually test it. This is an example of "dynamic similarity".

FICFB biomass gasification is a process for producing high caloric synthesis gas (syngas) from solid Hydrocarbons. The basic idea is to separate syngas from flue gas, and due to the separation we have a gasification zone for endothermic reactions and a riser for exothermic reactions. The bed material circulates between these two zones and serves as a heat carrier and a catalyst.

While researching the 250kW fluidized bed gasification pilot plant certain questions concerning particle dynamics in gas flows control arose. There is a zone where fluidized bed conditions are made with superheated steam, pneumatic transport with hot air and a pair of secondary gas inlets of CO_2. These particle flows are difficult to describe with mathematical models. This is the main reason why the three-time smaller cold-flow laboratory unit has been made. The hydrodynamics of particles will be studied in the air flow at arbitrary conditions. Flow conditions in the laboratory unit and pilot plant must be similar for a reliable evaluation of the process in the pilot plant.

Laboratory Unit

The laboratory unit is a device three times smaller than the pilot plant. Its main purpose is to simulate the hydrodynamic process of FICFB gasification in a cold flow. It is made from stainless steel and in the case of the parts that are of greatest interest to the present study is made of glass, so that the particle behaviour may be observed. Fig. 1 shows a model of laboratory unit. Its main elements are:

- Reactor (A),
- Riser (B),
- Cyclone (C),
- Siphon (D),
- Chute (E),
- Gas distributor (J_1 and J_2),
- Auxiliary inlets (I_1 and I_2).

Figure 1: Model of laboratory unit.

Firstly, let us look at the process. There are two gas distributors at the bottom of the reactor and riser, through which air is blown vertically. The pneumatic transport of the particles takes place in the riser, where they are separated from the air flow in cyclone and finally gathered in siphon. The second auxiliary inlet acts to fluidize the gathered particles and transport them to the reactor. Here, the fluidized bed is created with the upward blowing air. From here, the particles are transported to the riser through the chute and the speed of transportation is regulated by means of the first auxiliary inlet.

Table 1: Main dimensions of laboratory unit and pilot plant

	Laboratory unit	**Pilot plant**
$D_{gas,1}$ [mm]	100	300
$D_{gas,2}$ [mm]	190	600
D_{comb} [mm]	50	150
H_{comb} [mm]	1500	4500

We are primarily interested in how to establish a stationary and self-sustainable process. In the laboratory unit there are glass parts through which the process in course can be directly observed. However, in the hot flow model we will not be able to see what happens inside the pilot plant, and therefore our control system must be able to initiate the process, keep it in a stationary state and halt it on the basis of measured data such as relative pressure and flow velocities. For this mater, our laboratory unit consists of 7 pressure and 2 flow velocity measuring points. Fig. 2 details the positions of the pressure places.

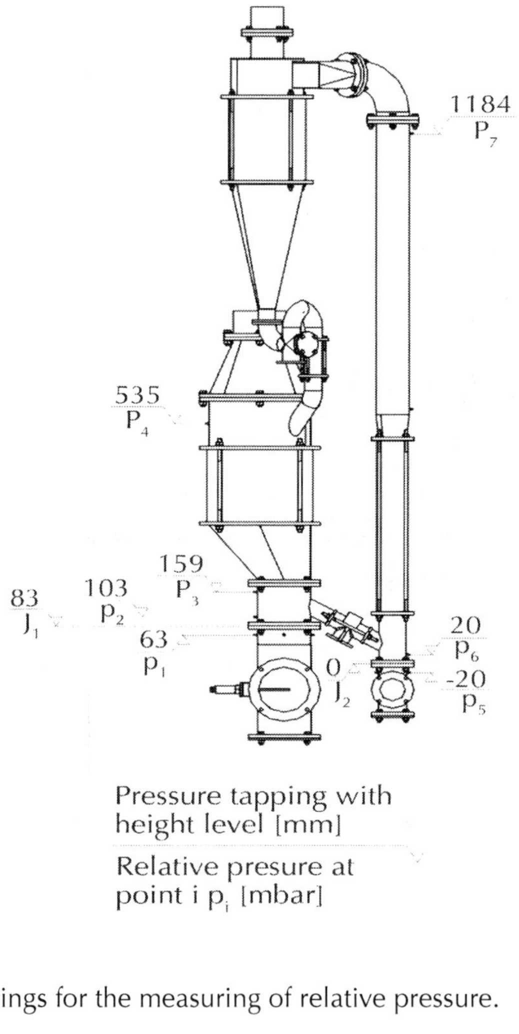

Figure 2: Openings for the measuring of relative pressure.

Trough experiments on the laboratory unit the effectiveness of elements will be studied so as to enable the correction and improvement of any construction flaws they contain. Fig. 3 shows the laboratory unit that will be used for studying the flow process. There are 7 places for pressure, 2 for temperature and 2 for gas flow measurements. For the proper operation of our solid flow system it is vital that the particles are maintained in dynamic suspension as settling down the particles can clog both the measuring openings and injection nozzles. Thus it is essential to design such systems with special care. All measurements involving the risk of clogging the measuring opening must be taken outside the solid flow zone if possible – gas flow velocity measurements with the Pitot tube must be taken in the gas pipeline before gas enters thru distributor. It is highly desirable for all measuring openings to be small and positioned rectangular to the direction of flow (Nicastro & Glicksman, 1982).

Figure 3: Laboratory unit.

Distributor

For the distributor 3 metal nets with openings of 225 μm have been used, with ceramic wool of 8mm placed in between as shown in fig. 4. We tried to achieve a sufficient pressure drop as to attain equal flow through the openings. According to Agarwal recommendation (Kunii & Levenspiel, 1991; Nicastro & Glicksman, 1982), the pressure drop across distributors must be 10 % of the pressure drop across the bed, with a minimum of 35 mm H_2O. With this we are in approximate agreement. At higher pressure drops across the distributor we get more particulate or smooth fluidization with less channelling, slugging and fluctuation in density. The pressure drop across the distributor is shown infig. 5.

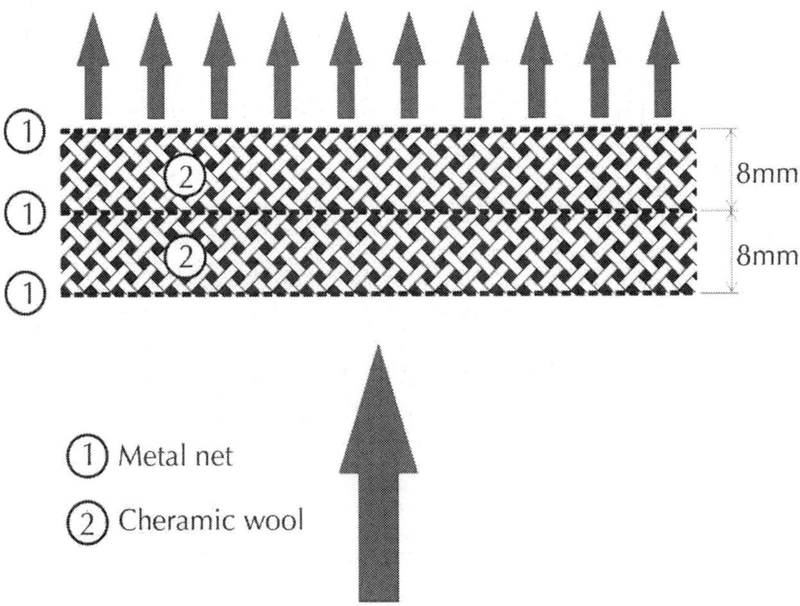

Figure 4: Distributor structure.

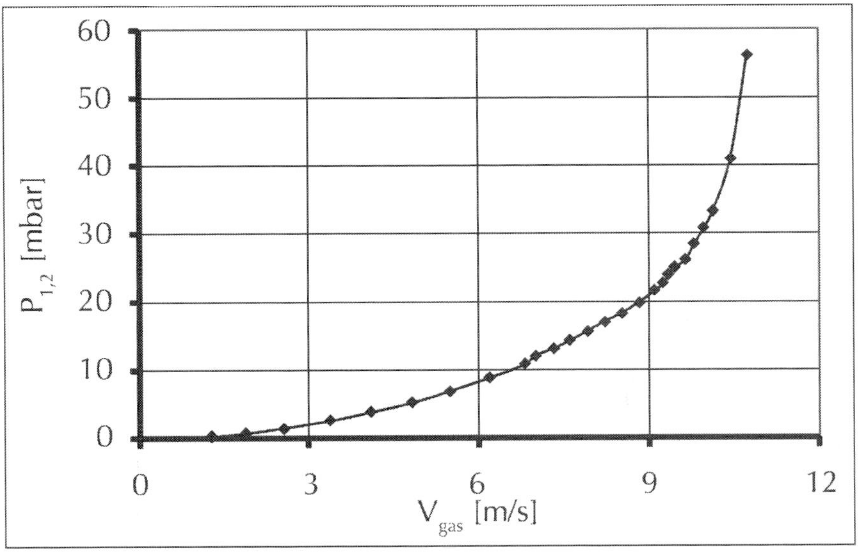

Figure 5: Pressure drop across the distributor with blowing of air.

Cyclone

In our case, the cyclone separator is placed behind the riser to separate the particles from the air flow. It has to be able to separate particles larger than 50 μm. For these conditions these particles are considered large as cyclones are often used for the removal of particles of about 10 μm diameter or larger from air streams. Our model is shown in fig. 6.

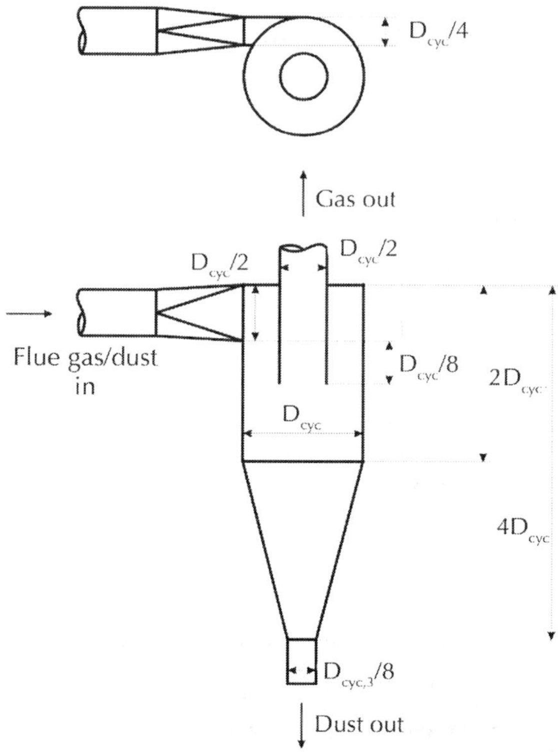

Figure 6: The characteristic dimensions of cyclone.

We dimensioned our cyclone according to Perry (Perry, 1988). $D_{p,50}$ is the particle size at which 50 % of solids of a given size are collected by the cyclone.

$$D_{P,50} = \sqrt{\frac{9.\eta_g.B_{cyc}}{\pi.N_s.v_g.(\rho_p - \rho_g)}} \tag{1}$$

By rearranging the equation (1), we obtain the following expression:

$$B_{cyc} = \frac{D_p,50^2 \cdot \pi \cdot N_s \cdot V_g \cdot (\rho_p - \rho_g)}{9 \cdot \eta_g} \tag{2}$$

The width of the cyclone entering the opening and the characteristic diameter are correlated by the following expression:

$$B_{cyc}=D_{cyc}/4 \tag{3}$$

The diameter of our cyclone is 150 mm. Smaller particles, which are not separated in cyclone are being collected in a filter placed on the cyclone gas exit.

BASIC EQUATIONS FOR DESCRIBING THE FLUIDIZED STATE AND SIMILARITY OF FLOWS

Reynolds Number

The goal herein is to compare flows in the laboratory unit to those in the pilot plant. In order for the two flows to be similar they must have the same geometry and equal Reynolds numbers. When comparing fluid behaviour at homologous points in a model and a full-scale flow, the following holds:

Re(laboratory unit) = Re(Scale-up pilot plant).

The Reynolds number of particles can be determined by the following equation (Kunii & Levenspiel, 1991):

$$R_{ep} = \frac{D_p \cdot v_g \cdot \rho_g}{\eta_g} \tag{4}$$

For achieving the required similarity, the following conditions must be also fulfilled:

$$\frac{p}{\rho_g \cdot v_g^2} = \frac{p_{g,ar}}{\rho_{g,ar} \cdot v_{g,ar}^2} \tag{5}$$

Minimal Fluidizing Velocity

The fluidization state starts when the drag force of by upward moving gas equals the weight of the particles (Oman, 2005)

$$F_{g-p} = \frac{1}{2} \cdot C_x \cdot A_p \cdot \rho_p \cdot v_g^{\ 2} \tag{6}$$

Or

$$\Delta p \cdot A_t = (A_t \cdot L_{mf})(1-\varepsilon_{mf})\left[(\rho_s - \rho_g)\frac{g}{gc}\right] \tag{7}$$

By rearranging equation (7), for minimum fluidizing conditions we find the following expression (Kunii & Levenspiel, 1991),

$$\frac{\Delta p_{mf}}{L_{mf}} = (1-\varepsilon_{mf})(\rho_s - \rho_g)\frac{g}{gc} \tag{8}$$

Voidage in fluidized bed $_{mf}$ is larger than in the packed bed and it can be estimated experimentally from a random ladling sample. For small particles and low Reynolds numbers the viscous energy losses predominate and the equation simplifies to (Kunii & Levenspiel, 1991):

$$v_{mf} = \frac{(\Phi_s \cdot D_p)^2}{150} \cdot \frac{\rho_p - \rho_g}{\eta_g} \cdot g \cdot \frac{\varepsilon_{mf}^{\ 2}}{(1-\varepsilon_{mf})} \tag{9}$$

for $Re_p < 20$

For large particles only the kinetic energy losses need to be considered:

$$v_{mf} = \sqrt{\frac{\Phi_s \cdot D_p}{1{,}75} \cdot \frac{(\rho_p - \rho_g)}{\eta_g} \cdot g \cdot \varepsilon_{mf}^{\ 3}} \tag{10}$$

For $Re_p > 1000$

If $_S$ and $_{mf}$ are unknown, the following modifications suggested by Wen and Yu (Kunii & Levenspiel, 1991) are used:

$$\frac{1-\varepsilon_{mf}}{\Phi S^2 \cdot \varepsilon mf^2} \cong 11 \tag{11}$$

$$\frac{1}{\Phi_S \cdot \varepsilon_{mf}^{\ 3}} \cong 14 \tag{12}$$

Equations (9) and (10) can now be simplified to:

$$v_{mf} = \frac{D_p^{\ 2} \cdot (\rho_p - \rho_g) \cdot g}{1650 \cdot \eta_g} \tag{13}$$

For $Re_p < 20$

$$v_{mf} = \sqrt{\frac{D_p \cdot g \cdot (\rho_p - \rho_g)}{24{,}5 \cdot \rho_g}} \tag{14}$$

For $Re_p > 1000$

Terminal Velocity

The upper limit of gas flow rate is approximated by the terminal (free fall) velocity of the particles, which can be estimated from the fluid mechanics (Kunii & Levenspiel, 1991):

$$v_t = \sqrt{\frac{4 \cdot g \cdot D_p \cdot (\rho_p - \rho_g)}{3 \cdot \rho_g \cdot C_x}} \tag{15}$$

There are spherical and non-spherical particle shapes in the bed and each of them has a different C_x value. If we combine equations (4) and (15) we get the velocity independent group:

$$C_x Re_p^2 = \frac{4 \cdot g \cdot D_p^3 \cdot \rho_g \cdot (\rho_p - \rho_g)}{3 \cdot \eta_g^2} \tag{16}$$

An alternative way of finding v_t for spherical particles uses analytical expressions for the drag coefficient C_x (Kunii & Levenspiel, 1991).

$$C_x = \frac{24}{Re_p} \text{ for } Re_p 0,4 \tag{17}$$

$$C_x = \frac{10}{\sqrt{Re_p}} \text{for } 0,4 Re_p 500 \tag{18}$$

$$Cx = 0,43 \text{ for } 500 Re_p 200000 \tag{19}$$

But still no simple expression can represent the experimental findings for the entire range of Reynolds numbers, so by replacing these values C_x in equation (16) we obtain:

$$v_t = \frac{(\rho_p - \rho_g) \cdot g \cdot Dp^2}{18 \cdot \eta_g} \tag{20}$$

For $Re_p < 0, 4$

$$v_t = \sqrt[x]{\frac{4}{225} \cdot \frac{(\rho p - \rho g)^2 \cdot g^2}{\eta g \cdot \rho g}} \cdot D_P \tag{21}$$

For $0, 4 < Re_p < 500$

And

$$v_t = \sqrt{\frac{3{,}1 \cdot D_p \cdot g \cdot (\rho_p - \rho_g)}{\rho_g}} \tag{22}$$

For $500 < Re_p < 200000$

Determining Density

In the pilot plant we will have multiple gas mixtures at different temperatures due to chemical reactions. For our calculations the density for these mixtures will be determined by the following equation (Oman et al., 2006):

$$\rho_g = \left(\sum_i \frac{wi}{\rho i} \right)^{-1} \tag{23}$$

To calculate the density of the gas mixture at an arbitrary temperature and an arbitrary pressure the density under normal condition must be calculated according to equation (24), with the obtained value being converted to density at the required parameters:

$$\rho_{g,ar} = \rho_g \cdot \frac{p_{g,ar}}{p_n} \cdot \frac{T_n}{T_{g,ar}} \tag{24}$$

Pressure Drops

With increased gas velocity of the small solid particles across the bed a characteristic state occurs. Pressure drop starts to increase, reaching its maximum value p_{mf} at minimum fluidization velocity v_{mf}. At this point only part of the bed is fluidized. When the bed is fully fluidized (at v_{mff}), the pressure drop is reduced to p_{mff} and is almost constant until gas reaches terminal velocity. If the velocity is still increasing, the particles start transporting pneumatically and pressure drop reduces rapidly to 0. By rearranging equation (8), we obtain the following expression (Kunii & Levenspiel, 1991):

$$\Delta p_{mf} = (1-\varepsilon_{mf})(\rho_s - \rho_g)\cdot g \cdot L_{mf} \quad (25)$$

The expression can also be extended to the fully fluidized state (Kaewklum & Kuprianov, 2008):

$$\Delta p_{mff} = (1-\varepsilon_{mff})(\rho_s - \rho_g)\cdot g \cdot L_{mff} \quad (26)$$

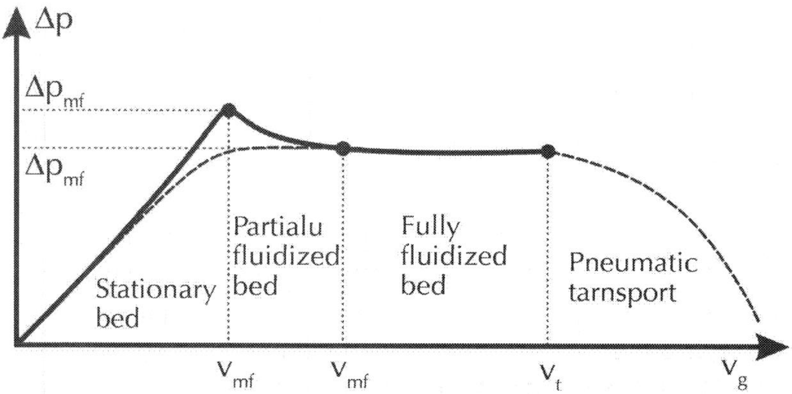

Figure 7: The change in pressure drop relative to gas velocity for Not-too-Small Uniformly Sized Particles (Kunii & Levenspiel, 1991).

A somewhat different differential pressure characteristic occurs with a wide size distribution of particles, which are usually present in industrial processes. When the gas velocity increases through the bed of solids, the smaller particles start to fluidize and slip into the void spaces between the larger particles, while the larger particles remain stationary (Kunii & Levenspiel, 1991) (see Fig. 8). However, after a full fluidization of bed material ($v_g > v_{mff}$), with increasing air velocity, pressure drop mainly remains constant.

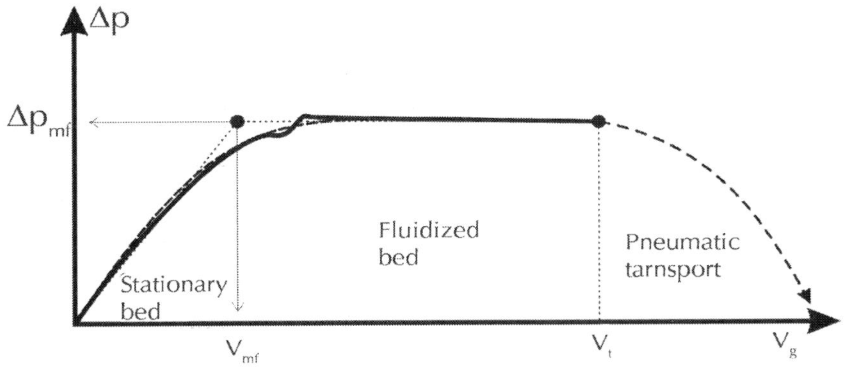

Figure 8: The change in pressure drop relative to gas velocity for Wide Size Distribution of Particles (Kunii & Levenspiel, 1991).

Mass Flows and Conservation Of Mass

For a regular flow process we have to ensure proper gas flows at the inlets. Through defining minimal fluidizing and terminal velocities, we can estimate the mass flow of the air reactor and riser by applying the following relations:

$$\varphi_{m\_g} = \frac{\rho_g \cdot v_g \cdot \pi \cdot D_{tube}^2}{4} \tag{27}$$

$$\varphi_{V\_g} = \frac{\varphi_{m\_g}}{\rho_g} \qquad q \tag{28}$$

It is extremely important to ensure that there are no mass losses between the ventilator and the reactor. It can be assumed:

$$v_g \cdot \frac{D_{gas},2^2 \cdot \pi}{4} = v_{g,ref} \cdot \frac{Dtube^2 \cdot \pi}{4} \tag{29}$$

CALCULATION ANALYSES

On the basis of the previously-mentioned equations, we can make an estimation of flow conditions in the reactor and riser. We have made a tabular comparison of physical properties between the laboratory unit and pilot plant in tables 2 and 3. The comparison is based on the established equality of Reynolds numbers. As mentioned in chapter 3.1. "In order for two flows to be similar they must have the same geometry and equal Reynolds numbers". In the laboratory unit, flows will be made with upward-blowing air at room temperature whereas in the pilot plant the fluid bed will be made with inlet of superheated steam and pneumatic transport with hot air blowing at 550° C.

Table2: Physical properties of gas in Reactor

	Reactor	
	Laboratory unit	**Pilot plant**
Gas	Air	Steam / Syngas
T [°C]	30	550 / 800
D_p [μm]	200	600
$_p$ [kg/m³]	8250	3025
$_g$ [kg/m³]	1,204	0,288 / 0,192
$_g$ [Pas]	1,8 10-5	3,1 10-5 / 4,6·10-5
$v_{Re<20}$ [m/s]	0,11	0,21 / 0,14
$v_{Re''/1000}$ [m/s]	0,75	1,58 / 1,95
$_m$ [kg/h]	6,4	158,9
$_V$ [m³/h]	5,4	548,5
R_ep	9,8	9,0 /4,9

In the meantime endothermic chemical reactions of pyrolisys, a water-gas-shift reaction will take place in the reactor while exothermic combustion occurs in the riser. Flue gases will have a the temperature of around 1000° C on exiting the combustor and syngas a temperature of approximately 800° C at the reactor's point of exit. Gases in the pilot plant will have lower densities and higher viscosities than the air in the laboratory unit. The bed material will be Olivine with D_p = 600 μm.

In order to establish similar conditions, we have to use smaller and denser particles. We have chosen brass particles with D_p= 200 μm. Simulation will also be tested with quartz sand and olivine.

Table3: Physical properties of gas in Riser

	Riser	
	Laboratory unit	**Pilot plant**
Gas	**Air**	**Air / Flue gas**
T_g [°C]	30	550 / 1000
D_p [μm]	200	600
$_p$ [kg/m^3]	8250	3025
$_g$ [kg/m^3]	1,204	0,61/0,294
$_g$ [Pas]	1,8 10-5	3,8 10-5/4,7 10-5
v_{Re}<0.4 [m/s]	10,1	15,7/12,6
$v_{0.4<Re<500}$ [m/s]	3,6	5,3/6,2
$v_{500<Re<200000}$ [m/s]	6,6	9,6/13,7
$_m$ [kg/h]	47,7	154,2
$_V$ [m^3/h]	39,6	524,5
R_ep	46,6	50,8 / 23,3

On the basis of studied flow velocities, mass flows, as well as pressure drops through air distributors and fluid beds at different points of the laboratory unit, we may anticipate the similar results in the pilot plant.

EXPERIMENTAL WORK

Firstly, we have to establish the fluidized bed in the reactor. The particles will fill the chute and the lower part of the riser. The chute is installed at the bottom of the reactor and riser and has an inclination angle. The fluidizing of the particles in the chute will then be started, along with the simultaneous initialization of the pneumatic transport of the particles. When sufficient material has been gathered in the siphon, the particles must be transported back to the reactor with the help of the first auxiliary inlet. The particles are now at their starting point. We must achieve a pressure at the bottom of the fluidized bed p_2 which is larger than that at the point where the chute connects to the riser p_6. The gas flow direction will be from the reactor to the riser, pushing the particles in the desired direction. At the top of the fluidized bed we have pressure p_4 which has to be lower than p_7, so the particles can now travel back to the reactor. But there has to be enough material in the siphon at all times in order to prevent the mixing of gases between the zones. Therefore, the siphon has to serve as seal gap for gases but not for material. The more gas goes through the siphon the lower the caloric value of the gas will be. Experiments will show how pressures are distributed across the system. Fig. 9shows which measured pressures are of greatest interest for our purposes.

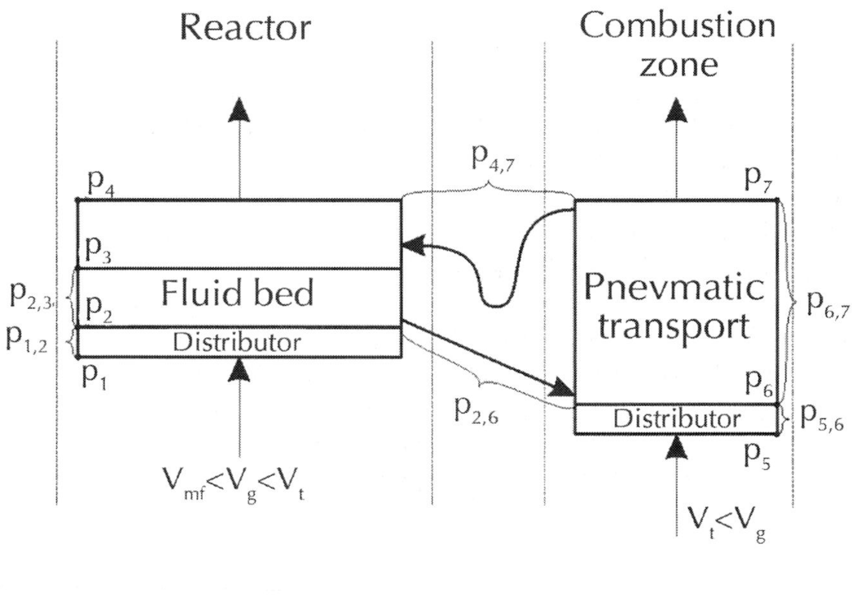

Figure 9: Measuring scheme.

By way of example, we will look at the experiment with quartz sand. The size of the particles used for simulation is shown in fig. 13. The particles have an average diameter of about 200 μm. A series of measurements were made and pressure drops at different bed heights taken. Fig. 10 represents a comparison of pressure drop across the bed in the reactor with the gas velocity for different bed heights.

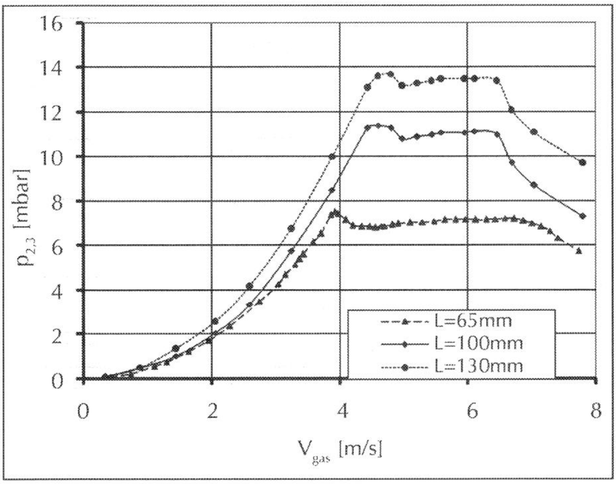

Figure10: Pressure drops over fluidized bed.

In lower beds less aggregative bubbling occurs and results closer to calculated values are obtained. Nevertheless, still there is a lot of deviation between them. In addition, there is some leakage of gas from the reactor through chute to the riser and as the Pitot tubes are placed in front of gas entering each zone those velocities do not represent the real situation, although the mass flow of air blown through unit is quite as predicted. However, gas velocity is almost impossible to measure within the laboratory unit because attempts to do so would inevitably lead to bed material clogging the measure openings in the device. Having said that, our assessment and purpose is to define and achieve a stationary process on the basis of the measuring system. The measured quantities are presented in table 4.

Table4: Measurements results

Symbol	**Value**	**[unit]**
p_1	34.4	mbar
p_2	11.3	mbar
p_3	0.2	mbar
p_4	0.1	mbar
p_5	6.2	mbar

$_{p}6$	3.9	mbar
$_{p}7$	3.2	mbar
$_{vga}s$	5.1	m/s
$_{vcom}b$	9	m/s

Comparisons of error between calculations and experimental results of pressure drops are presented in fig. 11 and 12. Through the application of the mathematical models we find that pressure drops can be predicted within a 20 % error margin. For example let us compare results between calculated and experimental values of pressure drop across 100 mm bed of quartz sand at minimum fluidization conditions. Calculating pressure drop according to equation 23 gives us 12.5 mbar, where physical properties are as follows: $_{p}$ = 2650 kg/m^3, $_{g}$ = 1.204 kg/m^3, $_{mf}$ = 0.55, L_{mf} = 110 mm ang g = 9,81 m/s^2. Bed height increases for 10 mm and so does consecutive voidage. A series of measurements gives us the average value for pressure drop which is $p_{2,3}$ = 11.4 mbar. As follows from this, the error of our prediction was 8.8 %.

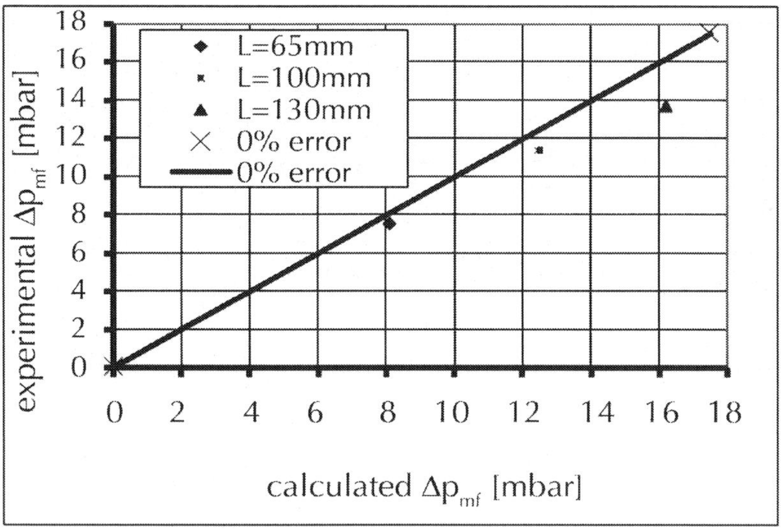

Figure11: The comparison of experimental and calculated p_{mf} for 200 μm quartz sand.

For calculating pressure drops across fully fluidized bed we use equation 26. The only difference comes with a little higher bed and voidage, which remain almost constant with increasing gas velocity to terminal velocity. So if we consider that L_{mff} = 115 mm and $_{mff}$ = 0.62 than pressure drop equals 11.4 mbar. With the comparison to the experimental value, which is 10.8 mbar, a 5.2 % error of prediction occurs. Error highly increases in aggregative and slugging mode of fluidization.

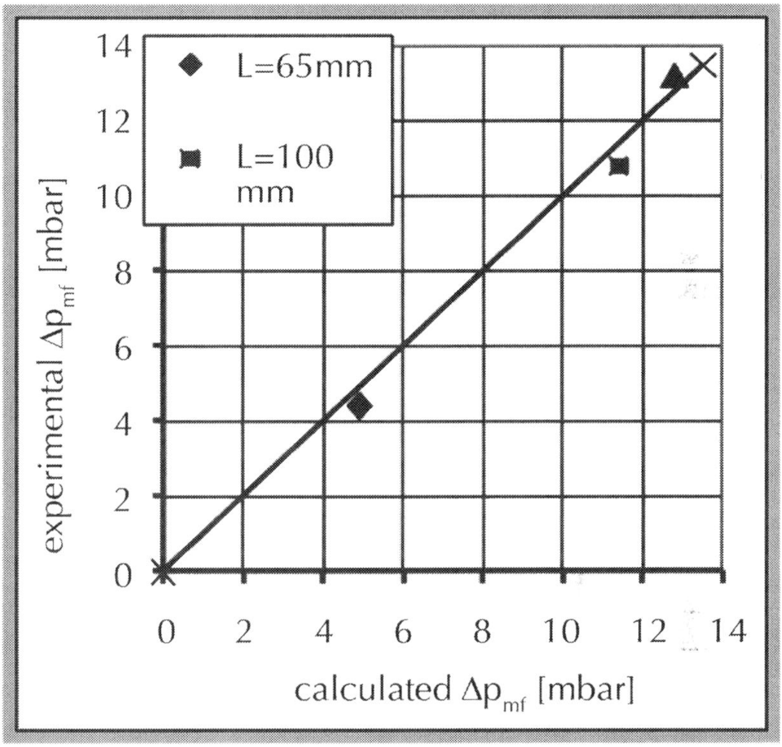

Figure 12: Comparison of experimental and calculated p_{mff} for 200 μm quartz sand.

Relative pressures were measured at a stationary state. One of the experiments was made when testing the process with quartz sand where the average particle diameter was about 200 μm. The stationary bed height in the reactor was 100 mm and the mass of sand used at simulation was 4.25 kg. When minimum fluidization conditions were obtained, the bed height increased by approximately 15 mm. A series of repeated measuring were carried out

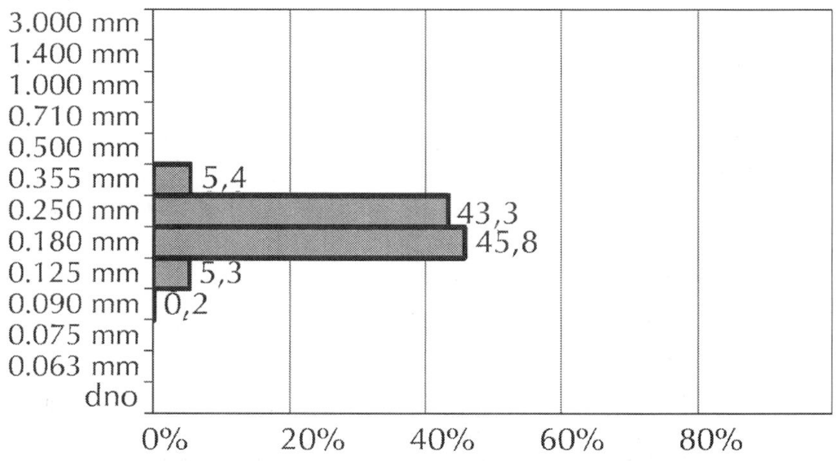

Figure13: The size of the particles used for simulation.

and the average relative pressure at the bottom of the fluid bed was p_2 = 11.3 mbar, with p_3 = 0.2 mbar the average value at the top. As follows from this, the pressure drop across fluidized bed was $p_{2,3}$= 11.1 mbar. Air flow had an average temperature of 25° C. Inlet gas velocity was about 5.1 m/s in the reactor and 9 m/s in the riser. We found a higher gas velocity for fluidization than calculated, due to a certain amount of air passing through the chute to the riser. This also provides the explanation as to why the measured terminal velocity in the riser was a little lower than anticipated, as the loss of air from the reactor helped increase the air speed in the riser – resulting in the aforementioned lower value.

COMPARISON TO THE PREVIOUSLY USED METHODS

Modern gasification is occurring in fluidized beds. Its advantage is using most fuels (wood, peat and coal) including agriculture "waste" such as straw, corn stover and manure. It has a potential to use municipal waste, such as garbage, it is quicker in response and it has shorter start up time. It lends itself to complete combustion applications which would allow it to use liquid wastes, such as used engine oil, non-recyclable plastics, junk mail & old shoes and garbage for the generation of heat. However, there is a problem of complex design. Still, nowadays most research efforts are being made on fluidization bed technology.

We tested a system very similar to the one tested by G. Löffler, S. Kaiser, K. Bosch, H. Hofbauer (Kaiser S. et al., 2003) with a minor difference. Our reactor had an eccentric diffuser which proved not to be a successful idea (Mele, J. et al., 2010). That is why in future research we are planning to test a reactor with a conical bed similar to those used by Kaewklum and Kuprianov. Our mathematical model is based on the derivation of Ergun's equation (Kunii & Levenspiel, 1991). L. Glicksman pointed out that for designing an accurate scale model of a given bed all of the independent non-dimensional parameters must be identical, such as considering the case where fluidized bed is operated at an elevated temperature of flue gas or at arbitrary conditions with air (Glicksman, 1982). Our work is also based on attaining similar non-dimensional parameters such as Reynolds and Euler numbers. The Freude number based on the minimum velocity, (v_{mf}/dp_g) has been proposed as the parameter to characterize the boundary between particulate and aggregative fluidization and the Archimedes number has been used to correlate a wide array of phenomena (Zabrodsky, 1966).

CONCLUSIONS

By observing the CFB processes in a three-time smaller laboratory unit with air flow the size and density of particles has been determined. The preferred option was to use brass powder with an average particle diameter of 200 μm. The assumption of particle flow similarity is based on a direct comparison of Reynolds numbers. In this case the Re_p are 9.8 and 9.0 in reactor and 46.6 and 50.8 in the riser. There is a 10 % difference between Re_p in both cases. Chemical reactions cause variations in temperature, density, and dynamic viscosity all of which affect Re_p. If we compare Re_p 9.8 and 4.9 at the reactor exit 46.6 and 23.3 at the top of the riser exit, we can see that Re_p changes by 50

% and the similarity at this point is actually questioned. By way of example, the experiment carried out with quartz sand was presented. When the process is stabilized and a smooth circulation is established, then pressure drops are as follows: $p_{2,3}$ = 11.2 mbar, $p_{6,7}$ = 0.7 mbar, $p_{2,6}$ = 7.4 mbar and $p_{4,7}$ = -3.1 mbar. This result set can be characterized as $p_2 > p_6$ and $p_4 < p_7$. Pressures are as expected and gas flows are in the appropriate directions. Through the application of the mathematical models we have, pressure drops can be predicted to within a 20% error margin. The experiments highlighted one major problem, namely that the cylindrical tube and asymmetric enlargement of the tube didn't prove to be a successful construction for the reactor. With beds higher than 13 cm fluidized beds are in aggregative or bubbling fluidization states. In turn, at bed heights over 30 cm even a slugging state is attained. The solution at this point is a conical bed design in accordance with Kaewklum and Kuprianov, 2008.

SYMBOLS

Table 5: Symbols

Ap	Cross-section of particle	[m^2]
A_t	Tube cross-section	[m^2]
B_{cyc}	Width of rectangular cyclone inlet dutch	[m]
c_x	Drag coefficient	
D_{cyc}	Characteristic cyclone diameter	[m]
D_{comb}	Riser diameter	[mm]
$D_{gas,1}$	Diameter of reactor upper segment	[mm]
$D_{gas,2}$	Diameter of reactor lower segment	[mm]
D_p	Diameter of particle	[μm]
$D_{p,50}$	Particle diameter at which 50% of particles are collected by cyclone	[μm]
D_{tube}	Inside tube diameter	[mm]
F_{gp}	Gravity of particle	[N]
g	Gravity acceleration [9,81 m/s2]	
g_c	Conversion factor [9,81gm m/s2 wt]	

${}_{Hcom}b$	Riser height	[mm]
i	Natural number	
j	Natural number	
L	Stationary bed height	[m]
${}_{Lm}f$	Bed height at minimum fluidization condition	[m]
${}_{Lmf}f$	Bed height at minimum fully fluidized state	[m]
${}_{N}s$	Number of turns made by gas stream in a cyclone separator	
p	Pressure	[Pa]
${}_{pg,a}r$	Pressure at arbitrary conditions	[Pa]
${}_{p}i$	Relative pressure in point i	[Pa]
${}_{pi}j$	Differential pressure between points i and j	[Pa]
${}_{p}j$	Relative pressure in point j	[Pa]
${}_{p}n$	Pressure at normal conditions	[Pa]
R_ep	Particle Reynolds number	
${}_{Tg,a}r$	Temperature at arbitrary conditions	[°C]
${}_{T}n$	Temperature at normal conditions	[°C]
${}_{vcom}b$	Gas velocity in riser	[m/s]
${}_{v}g$	Gas velocity	[m/s]
${}_{vg,re}f$	Gas velocity measured with pitot tube or orifice in tube before gas entering reactor	[m/s]
${}_{vga}s$	Gas velocity in gasification zone	[m/s]
${}_{vm}f$	Minimal fluidization velocity	[m/s]
${}_{vmf}f$	Minimal velocity of full fluidization	[m/s]
${}_{v}t$	Terminal velocity	[m/s]
p	differential pressure	[Pa]
${}_{pm}f$	differential pressure at minimum fluidization	[Pa]
${}_{pmf}f$	differential pressure at full fluidization	[Pa]
	Bed voidage	
${}_{m}f$	Bed voidage at minimum fluidization	
${}_{mf}f$	Bed voidage at full fluidization	
g	Dynamical viscosity of gas	[Pa·s]
${}_{g,a}r$	Dynamical viscosity of gas at arbitrary conditions	[Pa·s]

n	Dynamical viscosity of gas at normal conditions	[Pa·s]
g	Density of gas	[kg/ m^3]
p	Density of particle	[kg/ m^3]
m	Mass flow	[kg/h]
$_m$g	Mass flow of gas	[kg/h]
V	Volume flow	[m^3/h]
$_V$g	Volume flow of gas	

ACKNOWLEDGEMENTS

Investing in your future

OPERATION PART FINANCED BY THE EUROPEAN UNION

European Social Fund

Figure 14: "Investing in your future"- Operation Part Financed by the European Union – European Social Fund.

REFERENCES

1. L. R. Glicksman, 1982 Scaling Relationships For Fluidized Beds, Chemical engineering science, 39 1373 1384
2. R. Kaewklum, V. I. Kuprianov, 2008 Theoretical And Experimental Study On Hydrodynamic Characteristic Of Fluidization In Air-Sand Conical Beds, Chemical Engineering Science 63 1471
3. S. Kaiser, G. Löffler, K. Bosch, H. Hofbauer, 2003 Hydrodynamics of a Dual Fluidized Bed Gasifier- Part Ii: Simulation of Solid

Circulation Rate, Pressure Loop and Stability, Chemical Engineering Science, 58 4215 4223

4. D. Kunii, O. Levenspiel, 1991 Fluidization Engineering- Second edition, John Wiley & Sons, inc.,
5. G. Löffler, S. Kaiser, K. Bosch, H. Hofbauer, 2003 Hydrodynamics of a Dual Fluidized- Bed Gasifier- Part I : Simulation of a Riser With Gas Injection and Diffuser, Chemical Engineering Science, 58 4197 4213
6. J. Mele, J. Oman, J. . Krope, Jan, 2010 Scale-up of a cold flow model of FICFB biomass gasification process to an industrial pilot plant- hydrodynamics of particles, WSEAS transactions on fluid mechanics, 5 iss. 1, str. 15-24.
7. M. T. Nicastro, L. R. Glicksman, 1982 Experimental Verification of Scaling Relationships for Fluidized Beds, Chemical engineering science, 39 1373 1384
8. J. Oman, 2005 Generatorji Toplote, University in Ljubljana, Faculty of mechanical engineering, Ljubljana,
9. J. Oman, A. Senega nik, A. Mirandola, 2006 Air, Fuels and Flue Gases: Physical Properties and Combustion Constants, Edizioni Librerita Progeto, Padova, Italy
10. R. H. Perry, 1988 Perry's Chemical Engineers Handbook (6th ed.), New York: McGraw Hill International Ed.
11. S. S. Zabrodsky, 1966 Hydrodynamics And Heat Transfer In Fluidized Beds, The MIT press, Cambrige

Chapter 6

Second Law Analysis of Bubbling Fluidized Bed Gasifier for Biomass Gasification

B. Fakhim[1] and B. Farhanieh

[1]School of Mechanical Engineering, Division of Energy Conversion, Sharif University of Technology, Tehran, Iran

INTRODUCTION

The management of refused derived fuel (RDF) is one of the most significant problems especially for developing countries. Technologies to convert biomass energy already exist as well. Gasification through a bubbling fluidized bed gasified (BFBG) is discussed in this context. A BFBG is able to deal with wide variety of fuels due to the presence of inert bed material, in which bubbles mix turbulently under buoyancy force from a fluidizing agent like air or oxygen [1]. Under such violent bed conditions biomass waste particles are able to react fully to release volatiles as a result from high solids contact rate. Gases are released from the biomass particles and can then be used for producing electricity. In the literature there are several investigations on gasification processes from the thermodynamic point of view. Altafini and Mirandola [2] presented a coal gasification model by means of chemical equilibrium, minimizing the Gibbs free energy. The authors studied the effect of the ultimate analysis and the gasifying agents/ fuel ratio on the equilibrium temperature (adiabatic case) in order to obtain the producer gas composition and the conversion efficiency. They concluded that the equilibrium model fits the real process well. Similar conclusions for biomass gasification are presented by the same authors [3], simulating the gasifying process in a downdraft gasifier, where the object of study was the effect of the biomass moisture content on the final gas composition assuming chemical equilibrium. Lapuerta et al. [4] predicted the product gas composition as a function of the fuel/ air ratio by means of an equilibrium model. A kinetic model was used to establish the freezing temperature, which is used for equilibrium calculations in combination with the adiabatic flame temperature. The biomass gasification process was modeled by Zainal et al. [5] based on thermodynamic equilibrium. They analysed the influence of the moisture content and reaction temperature on the product gas composition and its calorific value. Ruggiero and Manfrida [6] emphasized the potential of the equilibrium model considering the Gibbs free energy. This proceeding can be used under different operating conditions for predicting producer gas composition and the corresponding heating value.

Many studies on the modeling of coal gasifers, in general, and coal gasification in bubbling fluidized beds, in particular, can be found in the literature. Nevertheless, thermodynamic modeling of the biomass gasification in bubbling fluidized beds has not been amply addressed. A few articles on the modeling of biomass bubbling fluidized bed gasifiers (BBFBGs) can be found in the literature. In modeling the biomass gasification (with air) in bubbling fluidized beds (BFBG), Belleville and Capart [7] developed an empirical model which was successfully applied to the biomass gasifier of Creusot Loire in Clamecy (France). Fan and Walawender [8] and Van den Aarsen [9] reported two of the pioneering models, which are well known today; Corella et al. [10] modeled some non-stationary states of BFBBGs; Bilodeau et al.

[11] considered axial variations of temperature and concentration and applied their results to a 50 kg/h pilot gasifier; Jiang and Morey [12,13] introduced new concepts in this modeling, especially related to the freeboard and the fuel feed rate; Hamel and Krumm [14] provided interesting axial profiles of temperature, although their work was mainly focused on gasification of coal and did not give many details of their model; Mansaray et al. [15,16] presented two models using the ASPEN PLUS process simulator.

In this work the equilibrium modeling of BFBG has been applied for the biomass waste gasification. The model employs equilibrium constants of all constituent reactions, in addition, the effect of the fuel/air ratio, moisture content of the fuel and gasifying temperature on the mole fraction of product gases of RDF gasification and corresponding higher heating value of it. Moreover, the exergetic efficiency and cold gas efficiency of the BFBG has been evaluated.

THE MODEL OF THE BFBG

Energy Analysis

The idealized fluidized bed gasifier model is used with the following assumptions:

1. The chemical equilibrium between gasifier products is reached,
2. The ashes are not considered and (iii) heat losses in the gasifier are neglected.

The global gasification reaction can be written as follows:

$$C_aH_bO_cN_dS_e+wH_2O+m\,(O_2+3.76N_2)\rightarrow n_1H_2+n_2CO+n_3CO_2+n_4H_2O+n_5CH_4+n_6N_2+n_7H_2S \quad (1)$$

In which the $C_aH_bO_cS_dN_e$ is the substitution fuel formula which can be calculated by the ultimate analysis of the fuel and the mass fractions of the carbon, hydrogen, oxygen, nitrogen and sulphur. "m" and "w" are the molar quantity of air entering the gasifier and moisture molar fraction in the fuel, respectively. The variable "m" corresponds to the molar quantity of air used during the gasifying process which is entering the BFBG at the temperature of 120° C and the pressure of 45 bar and depends on the gasification relative fuel/air ratio and the stoichiometric fuel/air ratio relating to the biomass waste as a fuel[17]

$$m=1/F_{rg}F_{st} \tag{2}$$

And w is determined from the moisture content of the fuel

$$w = M_{BM}\phi / MH_2O(1-\phi) \tag{3}$$

On the right-hand side, n_i are the numbers of mole of the species i that are unknown.

In a fluidized bed gasified, nearly the entire sulfur in the feed is converted to H_2S, which must be effectively removed to ensure that the sulfur content of the final gas is within acceptable limits. In the case of fluidized bed gasifiers, limestone can be fed into the gasified along with coal to capture most of the H2S produced within the bed itself. The limestone ($CaCO_3$) calcines inside the gasifier to produce lime (CaO), which in turn is converted to calcium sulfide (CaS) upon reaction with the H_2S inside the gasifier.

$$CaCO_3 \rightarrow CaO + CO_2 \tag{4}$$

$$CaO + H_2S \rightarrow CaS + H_2O \tag{5}$$

The substitution fuel formula $C_aH_bO_cS_dN_e$ can be calculated Starting from the ultimate analysis of the biomass waste and the mass fractions of the carbon, hydrogen, oxygen, nitrogen and sulphur (C, H, O, N, S), assuming a= 1, with the following expressions:

$$b = \frac{HM_C}{CM_O}, c = \frac{OM_C}{CM_O}, d = \frac{NM_C}{CM_N}, e = \frac{SM_C}{CM_S} \tag{6}$$

Ultimate analysis of the biomass waste (RDF) used in this model is shown in Table 1.

Table 1: Ultimate analysis of RDF (dry basis, weight Percentage) [18]

Waste Fuel	C%	H%	O%	N%	S%	Ash	HHV(MJ/Kg)
RDF	44.7	6.21	38.6	0.69	0.00	10.4	19.495

From the substitution fuel formula, the specific molecular weight of the biomass waste, the molar quantity of water per mole of biomass waste, the stoichiometric fuel/air ratio and the formation enthalpy of the biomass waste can be calculated.

Now for calculating the molar quantity of the product gases 7 equations are needed:

From the molar biomass waste composition $C_aH_bO_cS_dN_e$ and the molar moisture quantity, the atomic balances for C, H, O, N and S are obtained, respectively

C: $a=n_2+n_3+n_5$

H: $b+2w=2n_1+2n_4+4n_5$

O: $c+w+2m=2n_2+n_3+n_4$

N: $d+2m\times3.76=2n_6$

S: $e=n_7$

There are now only 5 equations to calculate 7variables. To solve the system, two other equations should be added. From the first assumption, two equations in equilibrium can be used. Chemical equilibrium is usually explained either by minimization of Gibbs free energy or by using an equilibrium constant. To minimize the Gibbs free energy, constrained optimization methods are often used which requires a realizing of complex mathematical theories. For that reason, the present thermodynamic model is developed based on the equilibrium constant. Therefore, the remaining two equations were obtained from the equilibrium constant of the reactions occurring in the gasification zone as shown below:

In the reduction zone of the gasifier, hydrogen is reduced to methane by carbon (methanation reaction).

$$C+2H_2\leftrightarrow CH_4 \tag{8}$$

Methane formation is preferred especially when the gasification products are to be used as a feedstock for other chemical process. It is also preferred in IGCC applications due to methane's high heating value.

The equilibrium constant K_1 relates the partial pressures of the reaction as follows:

$$k_1 = \frac{(P_{CH4}/P_{total})}{(PH_2/P_{total})} \quad (9)$$

Or as a function of the molar composition, assuming the behavior of the product gas to be ideal,

$$k_1 = \frac{n_5 \times n_{total}}{n_1^2} \quad (10)$$

The second reaction, also known as the water gas shift reaction, describes the equilibrium between CO and H_2 in the presence of water

$$CO+H_2O \leftrightarrow CO_2+H_2 \quad (11)$$

The heating value of hydrogen is higher than that of carbon monoxide. Therefore, the reduction of steam by carbon monoxide to produce hydrogen is a highly desirable reaction. The corresponding equilibrium K2 constant is obtained as follows:

$$k_2 = \frac{(P_{CO2}/P_{total})(P_{H2}/P_{total})}{(P_{CO}/P_{total})(P_{H_2O}/P_{total})} \quad (12)$$

Or as a function of the molar composition of the gas

$$k_2 = \frac{n_1 n_2}{n_2 n_4} \quad (13)$$

The values of the equilibrium constants K1 and K2 are calculated from the Gibbs free energy

$$K_p = \exp(-\Delta G^0_T/R_u T) \quad (14)$$

Where G^0_T is the difference of the Gibbs free energy between the products and the reactants:

$$G^0_T = H^0 - T\,S^0 \quad (15)$$

Substituting the Gibbs free energy in Eqs. (5) And (8), the equilibrium constants are obtained as

$$K_1 = \exp(-(G^0_{T,CH4} - 2G^0_{T,H2})/R_u T) \quad (16)$$

$$K_2 = \exp(-(G^0_{T,H2} + G^0_{T,CO2} - G^0_{T,CO} - G_{0T,H2O})/R_u T) \quad (17)$$

With

$$G^0T,\, i = \bar{h}^0_{f,\,298} + \int_{298}^{T} \overline{C_p}(T)dT - Ts^0 \quad (18)$$

Where $\overline{C_p}(T)$ the specific heat at constant is pressure in (J/mol K) and is a function of temperature. It can be defined by empirical equation below.

$$\overline{C_p}(T) = A + BT + CT^2 + DT^3 \quad (19)$$

In which the coefficients are obtained from the table 2

Table 2: Heat capacity of an ideal gas [19]

$\bar{C}_p(T)=A+BT+CT^2+DT^3$ (J/mol K)				
compound	A	**B×10²**	**C×10⁵**	**D×10⁸**
H_2	29.062	-0.82	0.199	0.0
O_2	25.594	13.251	-0.421	0.0
CO	26.537	7.683	-0.1172	0.0
CO_2	26.748	42.258	-1.425	0.0
CH_4	25.36	1.687	7.131	-4.084

Gasifying temperature

For calculating K1 and K2, the temperature in the gasification or reduction zone must be known. It should be noted that in bubbling fluidized bed the bed, temperature will be in the range of 900-1200° K by which the equilibrium constants will be calculated.

Enthalpy definition

After defining the corresponding equations, Because of nonlinear nature of some of the equations the Newton-Raphson method has been used to calculate the values n1-n7.

The enthalpy of the product gas is

$$\bar{h} = \sum_{i=prod} x_i \left(\bar{h}^0_{f,i} + \Delta\bar{h}_{T,i}\right) \qquad (20)$$

Where x_i is mole fraction of species i in the ideal gas mixture and $\bar{h}^0 f$ is the enthalpy of formation and $\bar{h}^T$ represents the enthalpy difference between any given state and at reference state. It can be approximated by

$$\Delta \bar{h}_T = \int_{298}^{T} \bar{C}_P(t)dT \quad (21)$$

Table 3 shows some the value of h¯⁰f for some gas components.

Table 3: Enthalpy of formation at the reference state [20]

Compound	h¯⁰f (kJ/mol)
H_2	0.0
O_2	0.0
CO	-110.52
CO_2	-393.51
CH_4	-74.85
$H_2O(l)$	-285.84
H_2S	-20.501[21]
SO_2	-296.833[21]

It should be noted that enthalpy of formation for solid fuel can be calculated as:

$$h_{f,bm} = HHV_{db} + \frac{1}{M_{bm}} \sum_{i=prod} v_i \tilde{h}_{f,i} \quad (22)$$

Where $(\bar{h}^0_f)$ k is the enthalpy of formation of the product k under the complete combustion of the solid and $\overline{HHV}$ is the higher heating value of the solid fuel.

Heat of formation of any biomass waste material can be calculated with good accuracy from the following equation [22]:

$$\Delta H_c = \overline{HHV}(KJ/Kmol) = 0.2326(146.58C + 56.878H - 51.53O - 6.58A + 29.45) \quad (23)$$

Where C, H, O and A are the mass fractions of carbon, hydrogen, oxygen and Ash, respectively in the dry biomass waste.

Energy Analysis

The entropy of ideal gas is represented by:

$$S = \bar{S}^0 + \int_{T_0}^{T} \frac{\overline{C_P}}{T} dT - R\ln\frac{P}{P_0} \qquad 24)$$

Where P is the pressure of the bubbling fluidized bed gasifier, and S^{--0} is entropy at reference state. Table 4 shows some components $S^{--}0$

Table 4: Entropy at the reference state (at T_{ref} =298.15K (25° C), p_{ref} =1 bar) [20]

Compound	S^{--0} (J/molK)
H_2	130.59
O_2	205.03
CO	197.91
CO_2	213.64
CH_4	186.19
$H_2O(l)$	69.94
H_2S	205.757[21]
SO_2	284.094[21]

The exergy of the product gas is comprised of two components: Exergy chemical exergy

(E^{CH}) and physical energy (E^{PH}).Total energy of the product gas is given as

$$E_{pg}=E^{PH}+E^{CH} \qquad (25)$$

The physical exergy is the maximum theoretical work obtainable as the system(here the product gas) passes from its initial state where the temperature is the gasifying temperature and the pressure equals the gasifier pressure to the restricted dead state where the temperature is T0 and the pressure is P0 and is given by the expression

$$E^{PH} = (H-H_o) - T_o (S-S_0) \quad (26)$$

The physical exergy of gas mixture per mole is derived from the conventional linear mixing rule

$$\bar{e}^{PH} = \sum x_i \bar{e}^{PH}_i \quad (27)$$

The chemical exergy is the maximum theoretical useful work obtainable as the system passes from the restricted dead state to the dead state where it is in complete equilibrium with the environment.

And chemical exergy of gas mixture is given by

$$\bar{e}^{CH} = \sum x_i \bar{e}^{CH}_{0,i} + RT_0 \sum_i x_i \ln x_i \quad (28)$$

Where $\bar{e}^{CH}_{0,I}$ is the standard chemical exergy of a pure chemical compound i which is available in Table 5 for some gas components.

Table 5: Standard chemical exergy of some substances at 298.15K and p_0 [21]

Substance	$\bar{e}^{CH}_{0,i}$ (kJ/kmol)
H_2	238490
CO	275430
CO_2	20140
$H_2O(g)$	11710
CH_4	836510
N_2	720
H_2S	812000[21]
SO_2	313.4[21]

Special considerations apply for the gasifying products when evaluating the chemical and physical exergy. When a product gas mixture is brought to P_0, T_0, some consideration would occur: At 25° C, 1 atm, the mixture consists of H_2, CO, CO_2, CH_4, N_2 together with saturated water vapor in equilibrium with saturated liquid. So it would be required to calculate the new composition at the dead state including the saturated liquid. Then the h¯° and s¯° values required to evaluate the physical exergy and the product gas mole fraction at the dead state essential for evaluating the chemical exergy can be calculated.

The exergy components and the total exergy for the moisture content of the fuel is obtained

$$E_{mois}^{PH} = w\left[\bar{h} - \bar{h}^{0} f_{,liq} - T_0(\bar{s} - \bar{s}^{0} H_2O_{(l)})\right] \quad (29)$$

$$E_{mois}^{PH} = w \times \bar{\varepsilon}_{0,H_2O_{(L)}}^{CH} \quad (30)$$

$$E_{mosis} = E_{mois}^{CH} + E_{mois}^{PH} \quad (31)$$

Energy for the fluidizing air entering the fluidized bed is defined with molar analysis of 0.21% O_2 and 0.79% N_2 with the pressure of 45 bar and the temperature of $373K_0$ by usinequations 25 and 26

$$E_{air} = E_{air}^{CH} + E_{air}^{PH} \quad (32)$$

For a biomass waste the chemical exergy is obtained as follows

$$_{0,\ biomass} = HHV_{biomass} \quad (33)$$

The factor is the ratio of the chemical exergy to the HHV of the organic fraction of biomass waste. This factor is calculated with the following correlation [18]:

$$\beta = \frac{1.0412 + 0.216(Z_H / Z_C\left[1 + 0.7884 Z_H / Z_C\right] + 0.045 Z_N / Z_C}{1 - 0.3035 Z_0 / Z_c} \quad (34)$$

Z_O, Z_C, Z_H and Z_N are the weight fractions of oxygen, Carbon, Hydrogen and Nitrogen, respectively in the biomass waste.

Therefore the total exergy of the biomass waste as a fuel can be defined:

$$E_{fuel} = \bar{}_{0,\ biomass} \times n_{fuel} \quad (35)$$

Heating Value and Efficiencies

Heating Value

The heating value of the producer gas can be obtained as the sum of the products of the molar fractions of each of the energetic gases (CO, H_2 and CH_4) with its corresponding heating value (Table 6).

TABLE 6: Heating value of combustible gases

LHV (MJ/kg mol)	HHV (MJ/kg mol)	gas
282.99	282.99	CO
241.83	285.84	H_2
802.34	890.36	CH_4
518.59	562.59	H_2S

Evaluation of the Efficiency

It is assumed that the fluidized bed gasifier operates as adiabatic and pseudo-homogeneous reactor at atmospheric pressure.

Gasification entails partial oxidation of the feedstock, so chemical energy of biomass waste is converted into chemical and thermal energy of product gas.

The first law thermodynamic or cold gas efficiency can be defined as the relation between the energy leaving the gasifier i.e. the energy content of the producer gas, and the energy entering the gasifier, i.e. the biomass waste and moisture. We assume the gas leaves the process at the reference temperature (25° C), losing the energy corresponding to its sensible enthalpy, and define the cold gas efficiency η_{Cg} as

$$n_{cg} = \frac{HHV_{gas}}{HHV_{biomas}} \quad (36)$$

Where HHV_{gas} and $HHV_{biomass}$ are the net heats of combustion (lower heating values) of gas and biomass waste, respectively.

The exergetic efficiency may be defined as the ratio between chemical exergy as well as physical exergy of product gas and the total exergy of the entering streams i.e. the biomass waste and the moisture and fluidizing air.

$$n_{Ex} = \frac{\dot{E}_{out}}{\dot{E}_{in}} = \frac{\dot{E}_{pg}}{\dot{E}_{air} + \dot{E}_{mois} + \dot{E}_{fuel}} \quad (37)$$

In this work variations of the exergy efficiency, cold gas efficiency and product gas concentration will be investigated as a function of temperature, gasifying fuel/air ratio (F_{rg}), and moisture content of the fuel ( ).

RESULTS AND DISCUSSION

Validation of the Model

The model presented in this article has been compared to the experimental work for the wood particles presented by Narvaez et al. [23]. By way of illustration the predicted HHV producer gas by the model and the results from the experiments are presented in Figure 1.

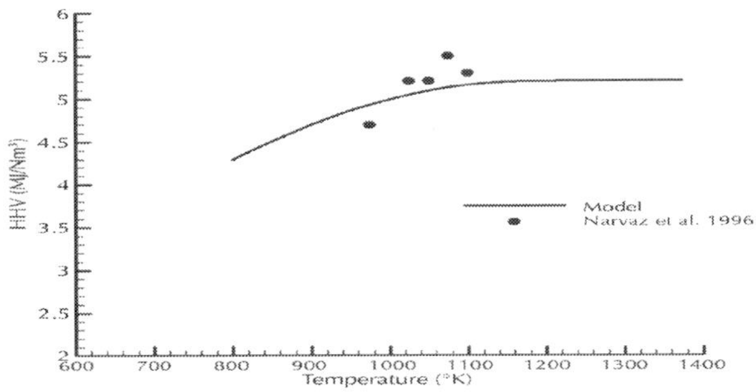

FIGURE 1: Higher heating values of product gas at different temperatures for wood particles.

Sensitivity Analyses

The effect of F_{rg} on product gas composition and higher heating value for RDF gasification is presented in Figure 2. An increase in Frg brings about an increase in the concentration of H_2 and CO and a substantial decrease in CO_2 concentration in dry gas product. This is because of the decreasing role of the char combustion in the bed compared to its gasification reaction, which results in higher concentration of combustible gases and lower CO_2.

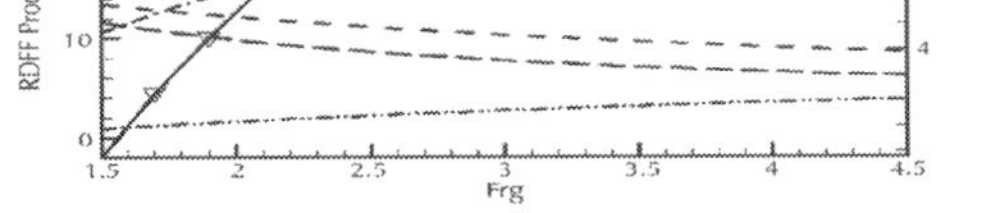

FIGURE 2: Concentration of product gases and higher heating value at different F_{rg} values and T_{bed} = 1100° K.

The effect of moisture content of the fuel on product gas composition and higher heating value for RDF gasification is presented in Figure 3.

As shown in the figure, an increase in moisture content brings about an increase in the concentration of H_2 and CH_4 and decrease in the concentration of CO. This is because of the increasing role of the moisture content of the fuel and effect of the methanation reaction (equation8) and the water-gas shift reaction (equation11) in which the molar concentration of the CO decreases because of the reaction with H_2O and production of H_2, ' and resulting an increase in the molar quantity of CH_4. Therefore the higher heating value will decrease as the moisture content increases.

The effect of gasifying temperature on product gas composition is shown in Figure 4. The figure shows that an increase in temperature brings about an increase in the concentration of H_2 and CO of RDF. This is because of the increasing role of the temperature in the equilibrium constants (16), (17) in which the equilibrium constant is dependent on the BFBG temperature, so an increase in temperature causes more production of combustible gases. The higher heating value in this temperature range at the constant F_{rg} is to some extent constant that is valid according to experimental works [22].

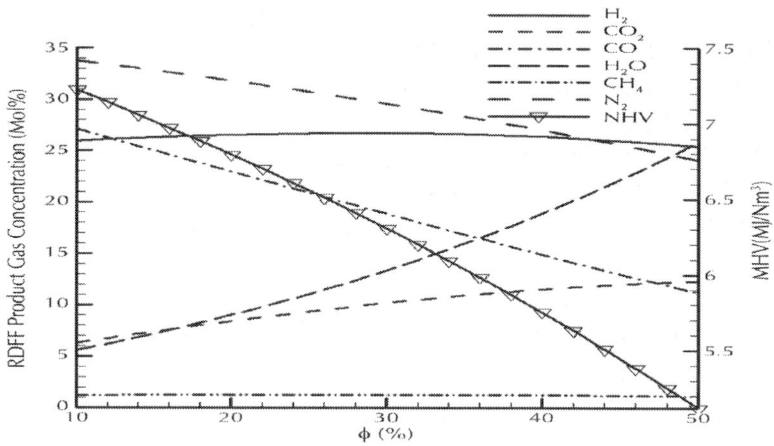

FIGURE 3: Concentration of product gases and higher heating value at different moisture content of the fuel at T_{bed} = 1100° K and F_{rg}=3.

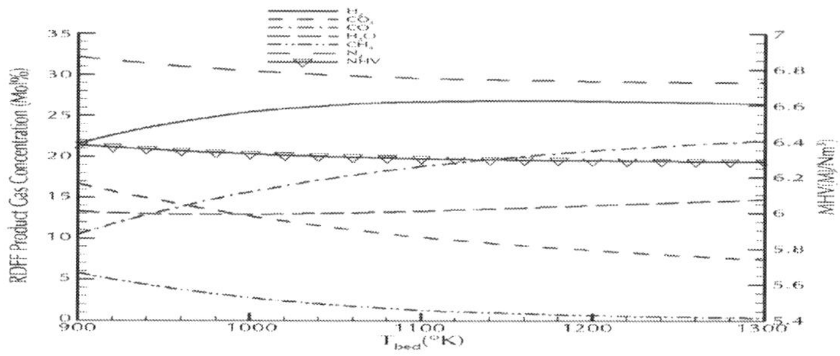

FIGURE 4: Concentration of product gases and higher heating value at various gasifying temperatures at F_{rg}=3.

The effect of F_{rg} with moisture content of the fuel on exergetic efficiency and cold gas efficiency for RDF gasification are presented (by line & flood contour type) in Figures 5, 6. It is shown that the exergetic efficiency of BFBG increases with rising fuel/air ratio because when less air is admitted to the process, the variations in mole fractions of product gases will influence the exergy of the product in comparison to exergy of the fuel. Higher moisture content will increase the exergetic efficiency because of its considerable effect on enthalpy of the product gases (figure 5). An increase in F_{rg}, as discussed before, brings about an increase in the concentration of combustible gases and higher heating value which yields an increase in cold gas efficiency and an increase in moisture content of the fuel, as discussed before, causes decrease in the concentration of combustible gases and higher heating value which yields a decrease in the cold gas efficiency (figure6).

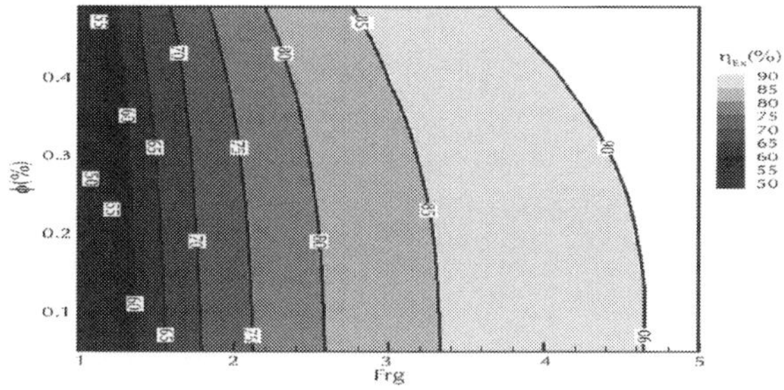

FIGURE 5: Exergetic efficiency of the gasifying process as a function of the gasifying relative fuel/air ratio and the moisture content.

The effect of F_{rg} and the bed temperature on exergetic efficiency and cold gas efficiency for RDF gasification are presented (by line & flood contour type) in Figures 7, 8. It is shown that the exergetic efficiency of BFG increases with rising fuel/air ratio as discussed for figures 5 and 6. Higher temperature will increase the exergetic efficiency because of its considerable effect on enthalpy of the product gases (figure 7). An increase in bed temperature, as discussed for figure 4, brings about an increase in the concentration of combustible gases and higher heating value which yields an increase in cold gas efficiency (figure8)

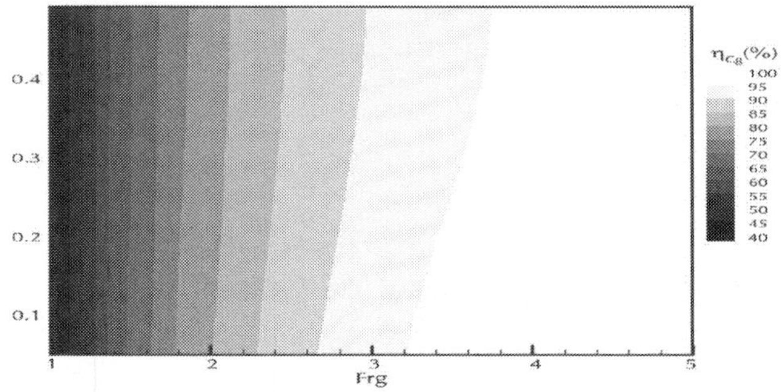

FIGURE 6: Cold gas efficiency of the FBG as a function of the gasifying relative fuel/air ratio and the moisture content.

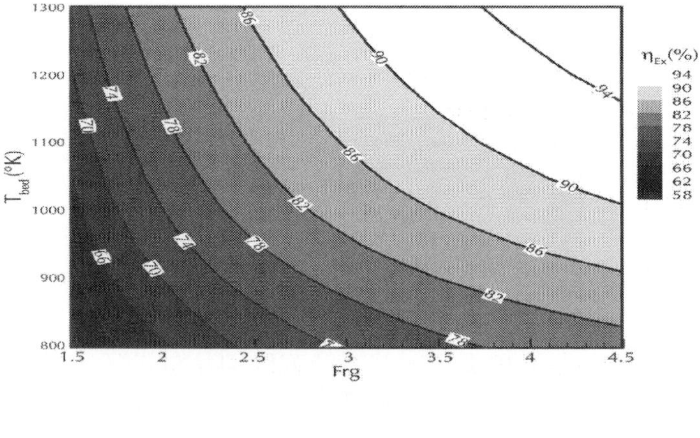

FIGURE 7: Exegetic efficiency of the gasifying process as a function of the gasifying relative fuel/air ratio and the gasifying temperature.

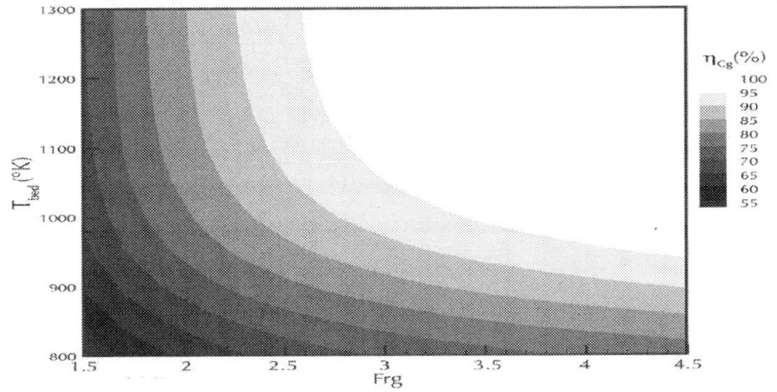

FIGURE 8: Cold gas efficiency of the FBG as a function of the gasifying relative fuel/air ratio and the bed temperature.

CONCLUSIONS

An equilibrium model was developed for the biomass waste gasification in the bubbling fluidized bed waste gasification. It was shown that higher moisture would decrease the product gas higher heating value as well as cold gas efficiency while increase the exergetic efficiency. Moreover, it was concluded that higher temperature and higher Frg would increase both the product gas higher heating value, cold gas efficiency and the exergetic efficiency.

NOMENCLATURE

TABLE 7: Nomenclature

C	mass fraction of carbon
H	mass fraction of hydrogen
$_{Fr}g$	gasification relative fuel/air ratio
$_{Fs}t$	stoichiometric biomass waste/air ratio
M	molecular weight (kg/mol)
MBM	biomass waste molecular weight (kg/mol)

N	mass fraction of nitrojen
m	molar quantity of air
w	molar quantity of biomass waste moisture content
$_{Ep}g$	Product gas total Exergy
$_{EP}H$	physical Exergy
$_{EC}H$	chemical Ecergy
$\Delta G_0 T$	gibbs free Energy((kJ/mol)
O	mass fraction of oxygen
$HH_{Vd}b$	higher heating value in dry base
P	pressure
$_s0$	standard Entropy(KJ/mol K)
S	mass fraction of sulphur
T	temperature

REFERENCES

1. Basu, P., Combustion and gasification in fluidized beds, Taylor & Fransis, 2006
2. C. R. Altafini, A. A. Mirandola, equilibrium. Chemical, of. Model, coal. The, process. Gasification, on. Based, minimization. The, the. Of, free. Gibbs, energy, In: Flowers 97, Florence, Italy, 1997
3. Altafini CR, Wander PR, Barreto RM. Prediction of working parameters of a wood waste gasifier through an equilibrium model. Energy Convers Manage 44 2763 77 2003
4. M. Lapuerta, J. Herna´ndez, F.V. Tinaut, A. Horillo, Thermochemical behaviour of producer gas from gasification of lignocellulosic biomass in SI engines. SAE Paper series 2001-01-3586, 2001
5. Z. A. Zainal, R. Ali, C. H. Lean, K. N. Seetharamu, of. Prediction, of. A. performance, gasifier. Downdraft, equilibrium. Using, for. Modeling, biomass. Different, materials, Energy Convers Manage 42 1499 515 2001 [6] Ruggiero M, Manfrida G. An equilibrium model for biomass gasification processes. Renew Energy 1999; 16:1106-9.

6. P. Belleville, R. Capart, A. model, predicting. For, gas. Outlet, from. A. concentration, gasifier. wood, A. V. in, Bridgwater (Ed.), Thermochemical Processing of Biomass, Butterworks, London, UK, 217 228 .
7. C.C. Chang, L.T. Fan, W.P. Walawender, Dynamic modeling of biomass gasification in a fluidized bed, AICHE Symposium Series 80234 (234) (1984) 80-90.
8. F.G. Van den Aarsen, Fluidized bed wood gasifier performance and modeling. PhD thesis, Univ. of Twente (NL), 1985
9. J. Corella, F. J. Alday, J. Herguido, A. model, the. For, states. Non-stationary, a. of, fluidized. Commercial, gasifier. Air, biomass. Of, G. in, Grassi, et al., (Eds.), Biomass for Energy and Industry, 2 Elsevier, London, 1990 2804 2809.
10. J., F. Bilodeau, N. The´rien, P. Proulx, S. Czernik, E. Chornet, A. mathematical, of. Model, biomass. Fluidized, The. Gasification, Journal. Canadian, Chemical. Of, Engineering, 1993 549.
11. H. Jiang, R. V. Morey, of. Pyrolysis, at. Corncobs, Biomass. Fluidization, Bioenergy, 31992 (1992) 81.
12. H. Jiang, R.V. Morey, A. numerical, of. Model, biomass. Fluidized, Biomass. gasifier, Bioenergy, 1992 (1992) 431.
13. S. Hamel, W. Krumm, Modelling, of. Simulation, fluidized. Bubbling, reactors. gasification, Proceed, 3rd European Conference on Fluidization, Toulouse, France, (2000), 29-31 May, 2000, 519 527 .
14. K. G. Mansaray, A. M. Al-Taweel, A. E. Ghaly, F. Hamdullahpur, V. I. Ugursal, modeling. Mathematical, a. of, rice. Fluidized, gasifier. Husk, I. Part-Model, Energy. Development, Sources, 2000 83.
15. K. G. Mansaray, A. M. Al-Taweel, A. E. Ghaly, F. Hamdullahpur, V. I. Ugursal, modeling. Mathematical, a. of, rice. Fluidized, gasifier. Husk, I. I. Part-Model, Energy. Sensitivity, Sources, 2000 167.
16. Melgar. Andre´s, F. Juan, Hannes. Pe´rez, Alfonso. Laget, Thermochemical. Horillo, modelling. Equilibrium, a. of, process. Gasifying, Conversion. Energy, Management, 2007 59-67
17. J. Krzysztof, Mark. J. Ptasinski, Anke. Prins, Exergetic. Pierik, of. Evaluation, gasification. biomass, Vol. Energy, 32, 568 574 , 2007

18. S. Jarungthammachote, A. Dutta, equilibrium. Thermodynamic, model, law. Second, of. A. analysis, waste. Downdraft, Energy. . gasifier, 2007 1660-1669
19. Yunus Cengel, Boles, Thermodynamics An Engineering Approach, 1999
20. Bijan. G. Adrian, Thermal. Tsatsaronis, Design, Optimization, 1996b by John Wiley & Sons.
21. S. A. Channiwala, P. P. Parikh, A. unified, for. Correlation, H. H. V. estimating, solid. Of, liquid, fuels. gaseous, Vol. Fuel, 81, 1051 1063 , 2007
22. I Narvaez, A. Orio, M. P. Aznar, J. Corella, gasification. Biomass, air. With, an. in, bubbling. Atmospheric, fluidized, Effect of six operationa variables on the quality of the produced raw gas, Ind Eng Chem Re,35 2110 2120 , 1996
23. Bubbling fluidized bed biomass gasification-Performance, process findings and energy analysis, Renewable Energy 33 2008 2008 2339 2343

Chapter 7

Harvest Systems and Analysis for Herbaceous Biomass

Jude Liu[1], Robert Grisso[2] and John Cundiff[2]

[1]Department of Agricultural and Biological Engineering, The Pennsylvania State University, University Park, Pennsylvania, USA

[2]Department of Biological Systems Engineering, Virginia Tech, Blacksburg, Virginia, USA

INTRODUCTION

Biomass is a distributed energy resource. It must be collected from production fields and accumulated at storage locations. Previous studies of herbaceous biomass as a feedstock for a bioenergy industry have found that the costs of harvesting feedstocks are a key cost component in the total logistics chain beginning with a crop on the field and ending with a stream of size-reduced material entering the biorefinery. Harvest of herbaceous biomass is seasonal and the window of harvest is limited. Biomass needs to be stored at a central location. Normally, several or many of these central storage locations in a certain range of a biorefinery are needed to ensure 24 hours a day and seven days a week supply. These centralized storage locations are commonly called satellite storage locations (SSL). The size and number of SSLs depend on the size of the biorefinery plant, availability of biomass within a given radius, window of harvest, and costs.

It is convenient to envision the entire biomass logistics chain from fields to biorefineries with three sections. The first section is identified as the "farmgate operations", which include crop production, harvest, delivery to a storage location, and possible preprocessing at the storage location. This section will be administrated by farm clientele with the potential for custom harvest contracts. The second section is the "highway hauling operations," and it envisions commercial hauling to transport the biomass from the SSL to the biorefinery in a cost effective manner. The third section is the "receiving facility operations," and it includes management of the feedstock at the biorefinery, control of inventory, and control of the commercial hauler contract holders to insure a uniform delivery of biomass for year-round operation.

Agricultural biomass has low bulk density, and it is normally densified in-field with balers, or chopped with a self-propelled forage harvester. Currently, there are four prominent harvesting technologies available for biomass harvesting [1]. They are: (1) round baling, (2) rectangular baling, (3) chopping with a forage harvester, and separate in-field hauling, and (4) a machine that chops and loads itself for in-field hauling (combined operation). Large round and large rectangular balers are two well-known and widely accessible harvesting technologies [2], which offer a range of advantages and disadvantages to farmgate operations. Round bales have the ability to shed water. When these bales are stored in ambient storage, they will store satisfactorily without covering and storage cost is significantly reduced. The round baler, because it is a smaller machine with fewer trafficability issues, can be used for more productive workdays during an extended harvest season over the winter months [3]. Large rectangular bales have greater bulk density, ease of transport, and increased baler productivity (Mg/h). However, the increased capital cost for the large rectangular baler and the bales' inability to shed water limit its use on farms in Southeastern

United States. Bale compression machines are available to compress a large rectangular bale and produce high-density packages [4]. The densified package has two or three times higher density than the field density of large rectangular bales [5].

The goal of an effective logistics system is to streamline storage, handling, and preserve the quality of the biomass through the entire logistics chain. This goal will minimize average feedstock cost across year-round operation. The farmer shares the goal to preserve the quality of the biomass, and also desires to produce the biomass at minimum cost. To assist in the accomplishment of the mutual objectives of both parties, this chapter will discuss major logistics and machine systems issues starting from the farmgate to the receiving facilities at a biorefinery. Constraints in this biomass supply chain will also be discussed. The impact of different harvest scenarios for herbaceous biomass harvest will be shown. Logistics systems have been designed for many agricultural and forest products industries. Thus, it is wise to use the lessons learned in these commercial examples. Each of these industries faces a given set of constraints (length of harvest season, density of feedstock production within a given radius, bulk density of raw material, various storage options, quality changes during storage, etc.), and the logistics system was designed accordingly. Typically, none of these systems can be adopted in its entirety for a bioenergy plant at a specific location, but the key principles in their design are directly applicable. Commercial examples will be used in this chapter to interpret these principles.

BIOMASS HARVESTING AND THE FIELD PERFORMANCE OF HARVEST MACHINE SYSTEMS

Introduction to Herbaceous Biomass Harvesting

Harvesting of cellulosic biomass, specifically herbaceous biomass, is done with a machine, or more typically a set of machines, that travel over the field and collect the biomass. These machines are designed with the traction required for off-road operation, thus they typically are not well suited for highway operation. Therefore, the transition point between "in-field hauling" and "highway hauling" is critical in the logistics system. In-field hauling is defined as the operations required to haul biomass from the point a load is created in-field to a storage location chosen to provide needed access for highway trucks. This hauling includes hauling in-field plus some limited travel over a public road to the storage location.

Harvesting systems can be categorized as coupled systems and uncoupled systems. Ideal coupled systems have a continuous flow of material from the field to the plant. An example is the wood harvest in the Southeast of the United States. Wood is harvested year-round and delivered directly to the processing plant. Uncoupled systems have various storage features in the logistics system.

Sugarcane harvesting is an example of a coupled system for herbaceous crops. The sugar cane harvester cuts the cane into billets about 38-cm long and conveys this material into a trailer traveling beside the harvester (Figure 1). The harvester has no on-board storage. Thus, a trailer has to be in place for it to continue to harvest. The trailer, when full, travels to a transfer point where it empties into a truck for highway hauling (Figure 2). Each operation is coupled to the operation upstream and downstream. It requires four tractors, trailers, and operators to keep one harvester operating. The trucks have to cycle on a tight schedule to keep the trailers moving. One breakdown delays the whole operation.

A "silage system" can be used to harvest high moisture herbaceous crops for bioenergy. With this system, a forage harvester chops the biomass into pieces about one inch (25.4 mm) in length and blows it into a wagon beside the harvester. This wagon delivers directly to a silo (storage location), if the field is close to the silo, or it dumps into a truck for a longer haul to the silo. All operations are coupled. That is, a wagon must be in place to keep the harvester moving, and a truck must be in place at the edge of the field to keep the wagons cycling back to the chopper. It is a challenge to keep all these operations coordinated.

A coupled system can work very efficiently when an industry is integrated like the sugarcane industry in South Florida, USA. Because the sugar mill owns the production fields surrounding the mill and the roads through these fields, the mill controls all operations (harvesting, hauling, and processing). Sugarcane has to be processed within 24 hours after harvest so the need for a tightly-controlled process is obvious.

Figure 1: Sugar cane harvester delivering material into a dump trailer for delivery to edge of field (Photo by Sam Cooper, courtesy of Sugar Journal, P.O. Box 19084, New Orleans, LA 70179).

Figure 2: Transfer of sugar cane from in-field hauling trailers to highway-hauling trucks.

An example of an uncoupled system is cotton production using the cotton harvester that bales cotton into 7.5-ft diameter by 8-ft long round bales of seed cotton. This system was developed to solve a limitation of the module system. With the module system, in-field hauling trailers (boll buggies), have to cycle continuously between the harvester and the module builder at the edge of the field. The best organized system can typically keep the harvester processing cotton only about 70% of the total field time. Harvesting time is lost when the harvester waits for a trailer to be positioned beside the harvester so the bin on the harvester can be dumped.

Baling is an uncoupled harvest system and this offers a significant advantage. Harvesting does not have to wait for in-field hauling. Round bales, which protect themselves from rain penetration, can be hauled the next day or the next week. Rectangular bales have to be hauled before they are rained on.

Field Capacity and Efficiency of Biomass Harvest Machines

The equipment used for baling and in-field hauling is a critical issue to the farm owners. More efficient harvest systems coupled with well-matched harvesting technologies specific to farm size and crop yield can minimize costs. The importance of understanding the linkage between various unit operations in the logistics chain was illustrated [6, 7]. Similarly, researchers have quantified the handling and storage costs for large square bales at a bioenergy plant [8]. However, in both of these evaluations the details of field operations and field capacities of machines involved in the field harvesting and handling were not available [9]. Instead, costs of bales at the farm gate were used to analyze bioenergy production costs. To maximize the field efficiency of field machine systems, it is essential for farm managers to know the field capacity of each machine involved in harvesting. In addition, quantitatively understanding the capacity of biomass harvest machines is essential to assess daily production and supply rate for a biorefinery or a storage facility.

The field efficiency of rectangular balers can be determined by calculating the theoretical material capacity of the baler and actual field capacity [10]. Calculation of material capacity can be demonstrated using a large rectangular baler as an example. The end dimensions of the large bales were 1.20 m × 0.90 m; bale length was 2.44 m. The depth and width of the chamber were 0.9 m and 1.2 m, respectively. Plunger speed was 42 strokes per minute. Measured bale density was 146 kg/m^3, and the thickness of each compressed slice in bale was 0.07 m. Thus, calculated theoretical bale capacity using equation 11.60 in [10] is 27.83 Mg/h. The plunger load could be set higher and produce higher density bales.

The theoretical capacity was obtained from a baler manufacturer under ideal conditions. Ideal conditions exist when a baling operation has [11]:

1. Long straight windrows
2. Windrows prepared with consistent and recommended density (mass/length)
3. Properly adjusted and functioning baler
4. Experienced operator

Actual field capacity of a baler will be impacted by the size and shape of the field, crop type, yield and moisture content of the crop at harvest, and windrow preparation. Typical field efficiencies and travel speeds can be found from ASAE Standards D497 [12]. Cundiff et al. [11] analyzed the field baler capacity and considered the effect of field size on baler field capacity. They found that the field capacities of round and large rectangular balers were 8.5 Mg/h and 14.4 Mg/h, respectively.

Another example of testing the baling capacity of a large rectangular would be the field tests conducted on wheat straw and switchgrass fields [13, 14]. Results showed that actual field capacity of a large rectangular baler was between 11 and 13 Mg/h. This indicates that the field capacity of a large rectangular baler could be 50% or less compared to its theoretical capacity.

Power Performance of Harvest Machine Systems

Machine system field efficiency is limited by tractor power performance, machine field capacity, and field conditions. Field conditions limit operational parameters and the percentage of the maximum available power. Since the high cost of harvest and in-field handling is still one of the main roadblocks of utilizing biomass feedstocks to produce biofuels, increasing machine system field efficiency through designing or selecting suitable machine systems is the challenge to machinery design and management professionals.

Power performance of a 2WD rear drive tractor was presented in a format of flow chart in ASAE standards [12, 15]. Total power required for a tractor is the sum of PTO power (*Ppto*), drawbar power (*Pdb*), hydraulic power (*Phyd*), and electric power (*Pel*) as expressed by equation (1). Depending on the type of implement, components in equation (1) may vary. Total power calculated with equation (2) is defined as equivalent PTO power, which can be used to estimate the tractor fuel consumption under specific field operations.

$$PT=PdbEmEt+ Ppto+Phyd+Pel \quad (1)$$

Where E_m and E_t are mechanical efficiency of the transmission and tractive efficiency, respectively. Each of the power requirements in Equation (1) can be estimated using recommended equations in [12]. Example of using this standard to estimate power requirements for a large rectangular baler system is available [13].

System Analysis Case Study - Large Rectangular Bale Handling Systems

Baling and Bale Collection

Baling, bale collecting, and bale storing of 650-ha wheat straw field at a commercial farm was studied as a typical large rectangular bale harvesting system. Factors that affect large rectangular bale production and handling logistics were quantified through observing complete field operations and then the system performance was analyzed and field capacities of all machines in this system was determined. System limitations were quantified, and means to reduce production costs were discussed [14].

Bales were stored in covered storage facilities. The equipment used included one large rectangular baler, two bale handlers, and three flatbed bale trucks (Figure 3). The straw was raked with a twin rotary rake. The rake was used to form uniform and evenly-spaced windrows from the straw that had been expelled by the combines. If the straw was rained on before baling, the rake fluffed the straw to enhance drying. The operator adjusted the swath of the rake in order to form windrows of proper size.

Figure 3: The large square baler, bale handler, and bale truck used in straw harvesting.

The field crew included a person operating the baler, two persons operating wheel loaders, and three truck drivers. The baler ran continuously throughout the day with the exception of operator breaks. Since no bales were left in the field overnight, the bales were collected at about the same pace as produced. A truck driver would bring the truck into the field and locate two bales. The wheel loader operator followed the truck through the field and loaded two bales at a time. When the truck was full, the driver would exit the field and transport the load to a storage facility. Bales were loaded in an interlocking manner, and the bales were not strapped down, a procedure that saved a lot of time.

Because there were three truck drivers and two trucks, a driver was not present while the truck was unloaded. The driver of the fully loaded truck would drive into the storage facility and position the truck to be unloaded. By the time this truck arrived with a full load of straw, the previous truck at the site would be empty. This process was repeated, with minor delays when a driver waited for a truck to be unloaded. Capacity of each operation was measured in terms of number of large rectangular bales (Table 1). Baling rate was the limiting factor in the system.

Bale Compression

To reduce long distance hauling cost of forage/hay bales, compression of these bales into ultra-dense bales was adopted by some producers to prepare the hay for overseas transport. Commercial bale compression equipment has potential to be used for the densification of baled biomass.

Table 1: Field capacities of large rectangular baler and bale handlers (Straw M.C., 14% w.b.) [14]

Field operation	Number of bales per hour (St. dev.)
Raking	85 (equivalent)
Baling	43 (9)
Accumulating into stacks of bales in field	93 (28)
Loading truck in field from accumulated bales	204 (29)
Loading truck in field from not first accumulated bales	143 (46)
Stacking bales in storage	93 (12)

Ten large rectangular switchgrass bales were compressed using the compressor (Figure 4) [9]. The input bale had a dimension of 0.86×1.22×2.21m, and the resulting compressed bales were 0.53×0.46×0.38m. When stacked in groups of 20 per pallet, the package had a dimension of 1.02×1.17×1.70m. The initial bales had a volume of 23.27m^3. When accounting for a 7.0% loss of material and water, this volume was 21.67m^3. The bales had average moisture of

14% (w.b.) at the time of compression. The compressed package had a volume of 13.13m^3, a reduction of 60.6% in volume.

Figure 4: Commercial large rectangular bale slicer/compressor.

System Analysis Case Study – Round Bale Handling System

The advantage of round bales is that the rounded top sheds water and bales can be stored in ambient storage without the expense of covered storage. A second important advantage of round bales is that round balers are conventional technology throughout the United States. They are widely used to harvest forage. Compared to a large rectangular baler, a round baler has lower capital cost.

A significant advantage is realized by the farmers if their round balers can be used for both their existing livestock enterprise and a bioenergy enterprise. Warm-season grasses are harvested during the winter for the bioenergy market. Thus, the biomass harvest does not conflict with the hay harvest. The advantage gained by the biorefinery is that no need for their feedstock producers to invest in new equipment. Requiring capitalization of new equipment will make the contacting with feedstock producers more difficult.

The disadvantage of round bales is that they are hard to stack and thus the handling and transport costs are higher. System performance was analyzed and field capacities were determined based on field measurements [16]. Individual handling of bales (either round or rectangular) is not cost effective. The high cost is caused by long loading and unloading times. Multi-bale handling units have been designed, and these units are discussed in Section 9 of this chapter.

FIELD WORKING DAYS – HARVESTING LIMITATIONS

Determining the schedule time required for harvest operations is an essential prerequisite to determine herbaceous biomass priority for securing inventory and effectively matching the transport of biomass from SSLs (satellite storage locations) to a biorefinery. Estimating the number of days expected to be available for baling is difficult because agricultural field work is heavily weather dependent and information concerning winter field operations is not well established. Harvest costs depend in part on the investment required in harvest machines (number of machines purchased), and this investment depends on the field capacity of the machines and the number of field workdays available during the harvest window. Therefore, an estimate of the number of harvest days is necessary to determine the total investment in harvest machines required to support a biorefinery.

The optimal perennial grass harvest strategy for a biorefinery remains to be determined. One potential strategy is to delay harvest until the plants standing in the field have transitioned from a non-dormant to a dormant state. Transitional events within the plant include translocation of nutrients from the above-ground plant parts to below-ground plant parts as initiation of degeneration (senescence) of above-ground tissues occur. Delaying harvest until the end of this transitional period, and harvesting over a short time period, can result in near maximum biomass yield, reduced moisture content, and reduced amount of nutrients removed with the biomass. This strategy would suggest an optimum harvest window for switchgrass from October to December. However, a harvest season this short would require substantial investment in harvest machines to complete harvest within a 3-month period. It also requires a large storage for inventory to supply the plant for year-around operation. By delaying harvest until nutrients have translocated, biomass tonnage will be decreased by this loss of mineral content, however since the relative amount of carbon in the material is increased, conversion efficiency and combustion quality may be improved [17]. Moreover, the translocated nutrients stored in the roots can be used for growth and development by the plant year after year, thus, reducing the need for and cost of supplementing the soil with nutrients through fertilization.

A second strategy is to extend harvest over as many months as possible. This would enable more economical use of harvest machines, reduce the quantity required in SSL storage, and potentially better match truck transport with SSL operations. With an extended harvest strategy, harvest would begin in July and extend through the following March. This extended harvest season would allow for spreading the fixed costs of the harvest machines over more Mg harvested per year thus reducing the $/Mg harvest cost. However, the earlier harvests before senescence (July through September) will remove more nutrients, and they will potentially conflict with other farm operations (i.e. grain/hay harvest, small grain seeding). To maintain productivity, additional fertilizer will be required, and this represents an additional cost to the farmer whose fields are harvested prior to October. The farmer whose fields are harvested January through March will have a lower yield because of leaf loss from the standing biomass exposed to winter conditions.

A delayed harvest raises an important issue relative to the payment to a farmgate contract holder. As previously explained, a certain loss occurs when the crop is left standing in the field for delayed harvest (December through March). Also, the delayed harvest increases the risk that a weather event will cause the crop to lodge so the harvest machine leaves material, and the "yield" is reduced. If these same fields were harvested during the optimum window (October through December) and stored longer in an SSL, there is a storage loss that increases with time in storage [18]. Considering the total reliance on stored biomass during the non-harvest months, which strategy is the better choice for the farmgate contract holder (i.e., generates the maximum quantity of biomass to sell to the biorefinery)? Is it better to delay harvest or build a larger SSL and invest in more harvesting machine capacity to complete the entire harvest within the 3-month optimum window? Epplin et al. [19] estimated differences in switchgrass harvest costs for both a 4-month and a 8-month harvest window. They accounted for differences in biomass yield across months, but they did not adjust for fertilizer requirements. They found that the estimated harvest costs varied from $25 per Mg for a 4-month harvest season to $11 per Mg for a 8-month harvest season. These results show that the length of the harvest season is a significant factor in determining the costs of harvesting biomass.

Epplin et al. [19] did not have refined estimates of the number of days that biomass could be harvested. They based their estimates of available harvest days on a study designed to determine the number of days that farmers in Southwestern Oklahoma of United States could conduct tillage operations. To determine more precise estimates of the number of harvest machines required to harvest, and thus obtain a more precise estimate of harvest costs, a more precise estimate of the number of harvest days is required.

Hwang et al. [20] determined the number of suitable field workdays in which switchgrass could be mowed and the number of days that mowed material can be baled. Empirical distributions of the days available for mowing and for baling switchgrass were determined for nine counties in Oklahoma. Distributions were determined for each month and for two potential harvest seasons (short, October–December and extended, July - February). Several conditions are necessary for safe baling of switchgrass. First, the soil must be sufficiently dry to support the weight of harvest machines. Second, prior to baling, the moisture content of the cut switchgrass must be at a level for safe storage. They provided evidence of the difference in harvest days across strategies.

Determining the time available for required harvest operations is a necessary, but not sufficient, prerequisite for determining the optimal harvest strategy. Additional information will also be required. As previously explained, if harvest begins in July prior to maturity and subsequent senescence in October, additional fertilizer will be required (the amounts are undefined) to offset the nutrients removed from the pre-senescence harvest. If harvest is delayed into December or later, the harvestable yield will be less (the amounts are undefined) than if it is harvested in October and November. Thus, the information regarding differences in harvest days across months must be combined with information regarding differences in fertilizer requirements, and biomass yields across harvest months, to determine an optimal harvest strategy.

CONSTRAINTS FOR A TYPICAL BIOMASS LOGISTICS SYSTEM

Constraints Set By Resource

Any biomass crop that provides an extended harvest season has an advantage because most biorefineries want to operate as many hours per year as possible. Continuous operation yields the maximum product per unit of capital investment. Herbaceous crops cannot be harvested during the growing season. Thus, storage is always a component in a single-feedstock logistics system design.

Two examples are discussed from opposite ends of the length-of-harvest-season spectrum. In the Southeastern United States, wood is harvested year-round. This is referred to as "stored on the stump." Weather conditions in the Southeast are such that few harvest days are lost in the winter due to ice and snow. On the other end of the spectrum, consider the harvest of corn stover in the Upper Midwest of the United States. The grain harvest is completed in the fall, and then the stover is collected. In some years, there are less than 15 days between the completion of the grain harvest and the time when the fields are covered with snow, and no stover can be collected. All feedstock required for year-round operation of a bioenergy plant, if it uses only corn stover, must be harvested in a three- to five-week period. This is a significant challenge for a logistics system.

Constraints Set By Purchaser of Raw Biomass

Biorefinery operators are interested in the cost of the feedstock as it enters their plant. The plant can burn the biomass directly to produce heat and power, or it can use the biomass as a feedstock for some more complex conversion processes to produce a high-value product. The term "feedstock" is used to refer to any raw biomass before its chemical structure is modified by a conversion process, which can be direct combustion, thermo-chemical, or biological.

Feedstock cost ($/dry ton) is defined as the cost of the stream of size-reduced material entering the reactor at the Bioenergy plant for 24/7 operation. The reactor is defined as the unit operation where an initial chemical change in the feedstock occurs. The reason for choosing this reference point for the biomass logistics system is two-fold.

1. The plants that can operate continuously, 24/7, have an advantage. Maximum production (tons/y) per unit of capital investment gives a competitive advantage in the market place – cost to produce the product is lower.
2. Some logistics systems do size reduction with the harvesting machine, while some size reduce at a transfer point between in-field hauling machines and highway hauling machines (perhaps as a pre-requisite to a densification step), and some size reduce at the entrance to the processing plant. In order to compare the several systems, it is necessary to have a consistent end point for the system analysis.

The readers should be aware that many studies in the literature select a different analysis endpoint than used here. A typical endpoint is the cost of feedstock when a truckload of raw biomass enters the plant gate. This reference point is favorable as many agricultural and forest industries pay the producer when the raw material is delivered to the plant.

FIELD LOSS

The yield losses of different herbaceous biomass feedstocks while standing in the field will impact the harvest systems and the windows of opportunities to gather the feedstocks into storage. The inventory losses will be impacted by herbaceous biomass feedstocks, package configurations and integrity, storage facilities and time in storage. Loss of dry matter is also an important parameter in biomass collection and transportation.

Biomass loses dry matter due to its high moisture or dryness. Leaves and other fragile parts of the plant are broken and lost in the wind or mixed with soil. Some of the losses occur during storage due to fermentation and breakdown of carbohydrates to carbon dioxide and other volatiles. Unfortunately the exact account of switchgrass losses in the field or during storage is not available. Sanderson et al. [21] reported dry matter loss in baling and storage of switchgrass and stated that the overall losses were less than for legume hay. They estimated that switchgrass bales stored outside without protection resulted in a dry matter loss of 13% of the original bale dry weight. They also estimated that dry matter loss of 1-5% during baling depending on the moisture content. Kumar and Sokhansanj estimated field and storage losses for straw and stover [22]. Other studies have also estimated the dry matter loss of biomass during storage, collection and transport [23-26]. Turhollow [27] estimated the losses from switchgrass to be similar to losses in alfalfa. The study estimated 8% losses for a mower-conditioner, 3% for a rake, 10% for a round baler, and 0.1% for a round bale wagon. He estimated average loss of 15% over 6 months of storage. A recent study showed that the dry matter loss in switchgrass collection (including storage) is less than 2% for different collection methods.

Moisture causes damage (microorganism growth) and subsequent dry matter losses in stored switchgrass bales. Several studies have shown that dry matter losses in switchgrass bales are greater for bales stored outside as compared to bales stored inside [18, 22, 28]. Moreover, dry matter losses are far greater for covered rectangular bales than uncovered round bales [18]. Large uncovered round bales had a better economic return than covered rectangular bales, when considering the cost due to mass loss during storage [18]. However, another study highlighted their successful use of rectangular bales [29]; the cost of covered storage was more than offset by the reduction in hauling cost for the square bales.

COMPUTER MODEL OF BIOMASS LOGISTICS SYSTEMS

When a series of operations are linked into a long chain of events, it is often difficult to see which operation is having the most impact on cost-effectiveness. Computer-assisted decision making tools are useful to select the most cost-effective logistics system. Recent studies [30-33] compared a number of different scenarios to determine the best unit operation options and identify bottlenecks. Mathematical programming approaches have been used to develop optimization models that are applicable to a variety of cases studies [34-37]. More recently, focus has shifted to simulation models using object-oriented programming [38] or discrete event simulation [39, 40]. The object-oriented approach [22, 41-43] simplifies scenario building, particularly related to various equipment options available for the same operation, since data pertaining to different options are stored in a standard object-oriented format.

The biomass logistics subject area can be divided into two types of modeling approaches depending on the environment encountered: stochastic or deterministic, and integer or continuous. In the areas of stochastic and deterministic environment, ISBAL and BioFeed capture the stochastic biomass logistics issues such as variability in processing as a result of weather, time, equipment breakdowns, etc. Additional work in this area can be found in [34; 40; 44-45]. In the deterministic environment, most work utilizes a geographic information system (GIS) interactively with optimization. This allows the GIS framework to work as a data management tool, which may call the optimization software directly. Most of the deterministic models are mixed-integer programs [37, 46-49]. One of the principles of designing an effective logistics system is starting with the feedstock resources distribution, and the simplest method is to utilize a GIS framework for data management.

Continuous models assume that all the land within a region is utilized for biomass production [50-52]. This approach uses average haul distances and cost. Hence, the solution obtained is difficult to implement due to lack of detail, as opposed to the solution obtained using an integer model formulation, where each production field is considered as a specific entity with specific costs, thereby resulting in precise decisions for implementation.

Modeling of Biomass Harvesting and Handling Systems for Field Operations

There are numerous studies developing for the optimum set of field equipment where the unit operations affect the performance of units upstream or downstream. Review of this literature is beyond the scope of this chapter. However, several simulation programs are highlighted that specifically addresses a biomass harvest and delivery system.

The Integrated Biomass Supply Analysis and Logistics Model (IBSAL) is a simulation model which simulates operations in the field

[22, 38, 41]. ISBAL is very useful for the simulation of operations that collects feedstock from the field and examines the flow of biomass into storage. IBSAL is best used for drawing conclusions about a sequential set of events. For example, if the user has a defined number of fields, a defined set of equipment, and a target number of tons to be harvested each month (each week) then IBSAL can provide valuable guidance for optimum biomass collection. The influence of weather, equipment breakdowns, and other disturbances to the biomass system can be "played" with a series of simulations.

The BioFeed model [43] determines the overall system optimum, and integrates the important operations in the biomass chain into a single framework. It is possible that the optimal solution recommended by BioFeed may not be implemented in a real system, either due to unforeseen disturbances such as weather or due to the actions of independent stakeholders such as farmers. However, an integrated model such as Bio-Feed determines the optimal configuration, system bottlenecks, and potential improvements. Such a model can be useful in quantifying the systemic impacts of technology improvement [42].

The BioFeed model results were compared to recent studies in literature [22, 31, 33]. The scope of the BioFeed model was similar to the scenarios developed in [22], where the delivered cost was estimated to be about $35/Mg (d.b.) excluding biomass size reduction. The major differences between ISBAL and BioFeed were harvesting and storage costs. While Kumar and Sokhansanj [22] ignored the storage costs, they also considered a self-propelled forage harvester which had a higher throughput capacity than the mower-conditioner considered in BioFeed, thereby reducing the cost. Khanna et al. [31] reported the switchgrass delivered cost of $64.84/Mg, which was similar to the BioFeed cost estimate. The study by Duffy [33] estimated the delivered cost to be $124.30/Mg. However, the major difference in the estimates was due to a much higher establishment cost.

Since the scope of the analyses and the assumptions differed for these studies, it is impossible to make specific comparisons. However, these comparisons illustrate that the overall model predictions agree reasonably well.

Modeling of Biomass Delivery Systems

Linear programming models have been used to analyze system interactions in biomass delivery systems. Dunnett et al. [53] proposed a program to optimize scheduling of a biomass supply system for direct combustion. This model simulated storage on farms and delivery to one location with a variable demand for heat. They suggest that costs of biomass handling can be improved 5 to 25% with the model recommendations. Bruglieri and Liberti proposed a "branch and bound" nonlinear model to determine biorefinery locations as well as the optimum transport method [54]. Their model focused on multiple feedstocks but did not use actual equipment performance data.

A model comprised for multiple purposes can bring attributes of benchmarking, simulation, and linear programming together to solve for the best solution. Leduc et al. a system of wood gasification plants optimized [55]. Their model focused on establishing a biorefinery plant in a location that is suitable for distribution of the product being manufactured (in this case, methanol). Other similar models have focused on silage handling operations [56].

A number of models have proven that a single chain of handling procedures can be optimized, but fail to adequately address the "disconnect" caused by storage, specifically satellite storage. Unfortunately, few models consider different harvest systems (or feedstock) supplying a single biorefinery. For example, one equipment system delivers chopped material directly from the field to the plant (probably from fields close to the plant), and a second set of equipment bales material and places it in storage which will be delivered during months when direct delivery of field chopped material is not possible.

As described earlier, a satellite storage location (SSL) is a pre-designated location that is used as a storage location for the biomass collected by the producers within a defined geographic region. SSLs are a logical transition point between "agricultural" and "industrial" operations and thus are critical elements in a logistics system design.

Study of Satellite Storage Locations (SSLS)

An SSL should be established where sufficient feedstock density will ensure the investment is justified. Location affects the costs associated with transporting the biomass from the production field to the SSLs and from the SSLs to the biorefinery. An additional factor that needs to be considered is how often a SSL is filled and unloaded. If a SSL is emptied only once a year, then the total SSL storage area would be doubled as compared to a twice-per-year unload schedule. Note, multiple fillings of a SSL is possible if harvest can be extended over several months and is properly matched with the SSL unload sequence.

Specialized equipment, with high productivity (ton/h) will be utilized to empty each SSL. This equipment can be permanent equipment for each SSL, or the equipment may be mobile and move from one SSL to the next. Questions that need to be answered to implement the mobile option are: 1) how many sets of equipment should be used to service the entire area? 2) What is the sequence of SSLs that each equipment set should unload? 3) Is an SSL ready to be unloaded? This last question directly affects the hauling company's contract. These questions are best considered in a virtual environment where scenarios can be compared and contrasted.

A case study to characterize feedstock resources and establish the SSL's was analyzed by Resop et al. [57] for a 48-km radius around Gretna, VA, USA. The GIS analysis identified potential production fields based on current land utilization determined using aerial photography and landuse classification (Figure 5). The analysis selected fields such that the production area was 6% of the total land area within the radius. The biomass produced is sufficient to supply the demand for a 1,944 Mg/d biorefinery, assuming 47 operating weeks per year. SSLs were established at 199 locations, and the existing road network (GIS database) was used to determine the travel distance from each SSL to the proposed plant location in Gretna. A weighted Mg-km parameter for transportation from the SSLs to the plant was computed to be 44.8 km, which implies that, averaged across all 199 SSLs, each Mg traveled an average of 44.8 km to the biorefinery.

Judd et al. studied three equipment options for the operations performed at a SSL [58]. Two options utilize the rack concept [59]. Ravula et al. analyzed a round bale logistics system utilizing this rack system; the rack size emulate a 20-ft (6.1-m) ISO container providing two levels of 8 bales, total of 16 bales, to be handled as a single unit [40]. A tractor-trailer truck can haul two racks (32 bales, 14.4 Mg). The results showed the rack hauling had higher transportation costs due to not loading the truck to maximum allowable weight.

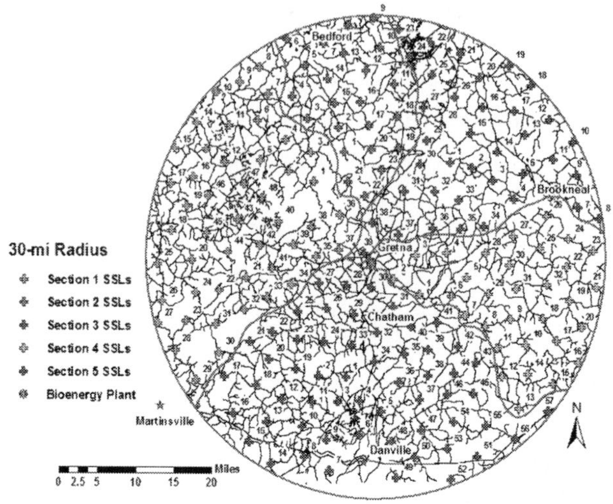

Figure 5: Example of Satellite Storage Locations (SSLs) located over a 30-mi (48-km) radius around a chosen bioenergy location. The refinery is located in the center of the circle. Each cross represents an SSL location with access to the public road system. The smallest SSL was a storage that can store biomass from 60 ac (24 ha) of production fields and the largest stores biomass from 1200 ac (486 ha) of production fields.

At a SSL, the round bales could be handled in one of two ways. The first option loads bales into the rack from the rear and is referred to as the "rear-loading." The second option, referred to as the "side-loading," loads the bales into a rack from the side. These two options were compared to grinding the bales at the SSL and compacting the ground material into a briquette (maximizes over-the-road load) for delivery [50]. Judd et al. [58] found that it was more cost effective to use the "side-load" option and haul the round bales than to form briquettes at the SSL, if the haul distance was less than 50 miles (81 km).

The large biorefinery concept calls for delivery of feedstock from a large geographic area. The results [58] suggest that some type of intermediate processing step, referred to as an Advanced Regional Biomass Processing Depot (ARBPD) by Eranki et al. [60], may be needed. These depots will convert the raw biomass to a more energy dense (higher value) product, and this product will then be delivered to a large biorefinery for final processing.

Why was the expense of size reduction and densification into briquettes at the SSL considered? Independent studies [61-63] reported that the tradeoff between the additional cost for densification and the reduction in transportation cost is significant for the hauling of logging residues. For additional work in the area of densification, see [64-65].

Application of Information Technologies

The inclusion of a hauling contractor in the business plan provides the best opportunity for all of the technology developed for other logistics systems to be applied to feedstock logistics. The information technologies applied include a GPS receiver in every truck and a bar code on every load as well as geographic mapping and routing management tools. Thus, time and location information can be observed at every SSL and the weigh-in scale at the receiving facility. The data collected can be used to optimize asset utilization in real time, and it will also feed needed data into the accounting software to compensate the farmgate and haul contracts.

It is expected that the collected data will be presented in real time. The information will provide a "Feedstock Manager," maps showing the location of all assets with updates at suitable time intervals. The goal is to provide an opportunity for the Feedstock Manager to make optimization decisions in real time. Examples are: trucks rerouted to avoid traffic delays, assets redeployed during breakdowns, increase at-plant inventory when inclement weather is forecasted, and a turn-down of plant consumption when a delay in feedstock deliveries cannot be avoided.

Some perspective of the logistics complexity, as presented to the Feedstock Manager, can be gained from the following sample parameters. This example presumes that operations will be in the Upper Southeast of the United States where switchgrass is harvested over an 8-month harvest season. Suppose the supply area has 199 SSLs within a 48-km radius of the plant, and each SSL has a different amount of material stored as shown in Figure 6. The producer desires to fill their SSLs at least twice during the harvest season to minimize per-Mg SSL investment. Suppose there are five SSL crews under contract and each of them wants the same opportunity to earn income (total Mg mass hauled per year). The Feedstock Manager's job is to treat all farmgate and haul contractors fairly.

Benchmarking Models for Biomass Logistics

It is appropriate to benchmark a proposed feedstock delivery system against existing commercial systems; for example, woodchips, grain, hay, sugarcane, and cotton. Ravula et al. [37] studied delivery to a cotton gin from over 2,100 field locations to better understand logistics system design for a biorefinery.

Switchgrass, corn stover, and other energy crops present two major challenges when benchmarked against other commercial examples. First, energy crops are spread over a greater number of sites when compared with woodchips harvested as byproducts from logging. Second, energy crops have low value when compared with hay, grain, sugarcane, or cotton. Applying benchmarked models to energy crop systems provides a starting point, but optimization techniques must be applied to "fine tune" the logistics system design.

Simulation of Mixed Harvesting Options for Supply of Single Plant

Brownell and Liu [66] developed a computer model to select the best option, or the best combination of options, for supply of a given bioenergy plant. The options include a direct chop and delivery option, two bale options (round and large rectangular), and a variation of the large rectangular bale option where these bales are compressed at a stationary location before shipment. Specifically, the options are: (1) loose biomass harvest using a self-loading forage wagon; (2) round bale harvest; (3) large rectangular bale harvest; and (4) large rectangular bale with compression before hauling.

The results of the model provide users a decision matrix which shows the optimized handling scenario for all four handling options analyzed by this program. Cost calculation was based on three plant demands (2,000, 5,000, and 10,000 Mg DM/d) to compare costs as the required production area expanded. Satellite storage locations were established based on the distribution of production fields and the existing road network.

The key constraints used for the simulation are:

1. Only loose biomass can be directly hauled from the field to the biorefinery.

2. Biomass cannot be stored at the plant, 20,000 tons (18,144 Mg) will "stand in field" in surrounding area of the plant and collected on demand as loose material.
3. Baled biomass will be transported to and stored at satellite storage locations.
4. Large rectangular bales may be compressed before transportation.
5. Biomass will be grown on one third of the available acreage, with an average yield of 3 Mg/ha (15% average moisture content).

The land close to the plant is more valuable, from the biorefinery viewpoint, for the production of feedstock. The study found that a plant can afford to pay more for feedstock (landowner gets a higher price) if a farm happens to be closer to the chosen plant location.

The results showed that the size of SSLs is sensitive to the amount of material demanded by the plant. The program tests various sizes, distances between SSLs, and possible overlaps of production fields "served" by each SSL. The programming optimizes the simulation by slowly changing the size and location of SSLs. The program solves for the maximum amount of acreage available in the simulation model. The output is a distance material in each SSL can be efficiently hauled based on if it is harvested and hauled directly from the field or if it is baled and stored before hauling. The program also solves for the lowest cost for a set acreage when the closest fields are field chopped and remaining fields are baled.

COMMERCIAL EXAMPLES OF LOGISTICS SYSTEMS

Typical Linkage in Logistics Chain

As previously stated, biomass logistics is "a series of unit operations that begin with biomass standing in the field and ends with a stream of size-reduced material entering a bioenergy plant for 24/7 operation." This concept can be visualized as a chain (Figure 6), where the links are unit operations. The example in Figure 6 shows a logistics system moving bales from the field to SSLs, and then from these SSLs to a bioenergy plant. The dotted lines show various segments of the chain that are assigned to the several entities in the business plan. In this example, the segment identified "farmgate" is performed by the feedstock producer, the segment labeled "SSL" is performed by a load-haul contractor; the segment identified as "Receiving Facility" is performed by the bioenergy plant. Figure 6 is one example; several other options are used in commercial practice. In next section, three commercial examples are given to show readers the range of options in a logistics system.

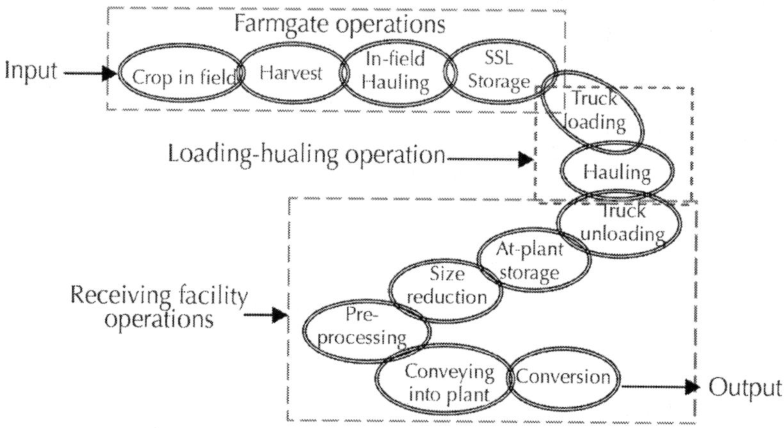

Figure 6: Logistics chain for delivery of round bales to a bioenergy plant (The dotted lines identify segments of the chain which are performed by different business entities).

Examples of Commercial Logistics Models

Traditional Model

The producer grows, harvests, stores, and delivers raw material (biomass) in accordance with a contract with the processing plant. Deliveries are made to insure the plant has a supply for continuous operation during the processing season. For many agricultural industries (i.e. cotton, grain, sugar cane, fruits and vegetables), the processing season approximately coincides with harvest season, and it is only part of the year. The advantage of the traditional model, from a processor's point of view, is that all quality issues reside with the producer. There is no question who is responsible if a quality standards are not met. Business people planning the operation of a bioenergy plant tend to prefer this model, though they typically balk at paying a feedstock price that adequately compensates the producer for their additional risk.

Cotton Model

A cotton producer grows, harvests, and stores the raw material (seed cotton) in modules at an edge-of-field location with suitable access for highway hauling trucks (Figure 7). The gin (processing plant) operates a fleet of trucks to transport the modules as required for operations during the ginning season. Farmers are paid for the seed cotton that crosses the scale at the gin. The gin operates a warehouse and stores bales of ginned cotton for periodic delivery to provide a year-round supply for their textile mill customers.

Figure 7: Module hauler picking up module stored at edge of field.

The advantage of the cotton model is that the specialized equipment used for hauling the modules is owned, scheduled, and managed by the gin. Thus, the producer does not own a piece of equipment they will use only a few times a year. The gin deploys the module haulers to their customers, thus the haulers accumulate more hours per years over what a producer would use, and the hauling cost ($/Mg) is minimized [40]. Typically, gins haul modules in the order they are "called in" by the producers. Module storage time in the field, and any subsequent losses (at the gin), are not dealt with in the producer-gin contract. The producer is paid the contract price for the mass of cotton fiber (and co-products) that the gin produces from a particular module.

Sugarcane (Texas Model)

The producer grows the sugarcane. The sugar mill takes ownership of the crop in the field and harvests and transports it for continuous operation during the processing season. Sugarcane must be processed within 24 hours of harvest so storage is not an option within the sugarcane system.

The advantage of the Texas model is that the producer does not own any harvesting or hauling equipment. The disadvantage of the system is the producer does not have control over the time of harvest. Sugar content peaks (point of maximum compensation) in the middle of the season and a producer who has their cane harvested early, or late, sells less sugar. Fortunately, the issue is addressed in their contract.

DEVELOPMENT OF CONCEPT FOR MULTI-BALE HANDLING UNIT

Modulization of Bales

Individual handling of bales (round or square) is not cost effective because of long loading and unloading times. Several concepts for a multi-bale handling unit are being developed. A verified multi-bale system is not available at this time to use as an example.

Permission was received [67] to present a concept that is far enough along in development that some field tests have been done. The concept was developed by a consortium led by FDC Enterprises and is shown in Figure 8. The self-loading trailer loads six stacks of six large rectangular bales, referred to "6-packs," for a total load of 36 bales. The bale size is 0.91 m × 1.22 m × 2.44 m. The length of the load is 6 × 2.44 = 14.6 m (48 ft), the height is 3 × 0.91 = 2.74 m (9 ft), and the width is 2 × 1.22 m = 2.44 m (8 ft).

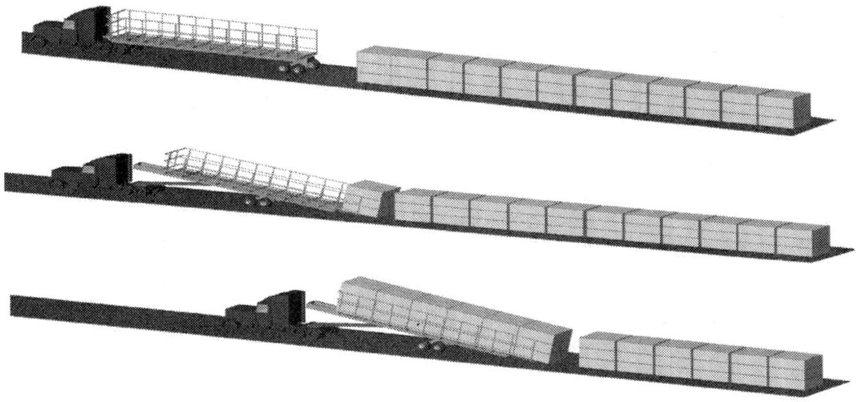

Figure 8: Multi-bale handling unit concept developed for 3x4x8 large rectangular bales by FDC Enterprises-led Consortium (Reprinted with permission [67]).

The trailer built to implement the concept [68] is shown in Figure 9. Estimated load time is 5 min, which is about the same load time for the cotton module (Figure 7). The 36-bale unit can be off-loaded by the truck directly onto the conveyor into a bioenergy plant, or it can be off-loaded into at-plant storage to be used later, just as is done with cotton modules at a cotton gin.

A similar concept known as the "Rack System" envisions that round bales will be loaded into a rack in-field or at an SSL. This rack is lifted off a trailer at the plant, emptied as the bales are needed, and then returned to be refilled. The racks are cycled multiple times each week within the closed logistics system. Cundiff et al. [59] developed the rack system concept, and it will now be used as an example to illustrate how the principles required for a logistics system are implemented.

All the various concepts and options cannot be discussed. However, we will apply a specific example to help the reader think through a process of developing a logistics system. The selection of the "Rack System" for this example implies no criticism of other ways of implementing the multi-bale handling unit concept currently being developed.

Figure 9: Self-loading trailer built by Kelderman Mfg. to implement multi-bale handling unit concept for 36-bale stack of large rectangular bales.

Receiving Facility

When a truckload of feedstock arrives at the bioenergy plant, what happens? Does this material go directly into the plant as in a just-in-time processing model? Or, does it go into some type of at-plant storage? If it goes into at-plant storage, is there an efficient procedure for placing it into storage and retrieving it? All these questions relate directly to the cost of Receiving Facility operations, thus, when a bioenergy plant commissions a logistics system, it first specifies how it wants to receive material. Key questions for a receiving facility are:

1. Do I want each load to be the same size with the truck loaded the same way (i.e. are bales need to be in the same configuration)?
2. What are the hours of operation, and can I schedule approximately the same number of deliveries each hour of the workday?
3. What size at-plant storage must be maintained?

Design of a receiving facility is beyond the scope of this chapter; but, it is hoped that by posing these questions, the reader will begin to think about the constraints that a bioenergy may have when specifying a logistic system. Two criteria used in this example are:

1. Weigh and unload a truck in 10 min, and
2. Cost effective flow of material in-and-out of at-plant storage to support 24/7 operation.

Design of a logistics system is directly linked to the design of the receiving facility – neither is designed independent of the other.

Farmgate Contract

Creation of a multi-bale handling unit will require specialized equipment. It is not a cost-effective to require each farmgate contract holder to own this equipment The rack system, for example, envisions a farmgate contract whereby the contract holder grows the crop, harvests in round bales, and places these bales in storage at a specified SSL. The contract holder owns and maintains the SSL and is paid a storage fee for each unit of feedstock that is stored. The biomass is purchased by the bioenergy plant in the SSL. All agricultural operations are now "sequestered" in the farmgate contract which gives those seeking a farmgate contract a well-defined process to prepare their bid.

A SSL must be graded to a minimum slope specification, must be near a main road, and must have a compacted gravel base. The gravel reduces bale degradation from water damage on the bottom of the bale and provides a suitable surface for equipment operations. Each SSL will receive material from either a single large production field or multiple, smaller fields as proposed by Cundiff et al. [59].

Hauling Contract

The rack system envisions that the hauling contractor will invest in industrial equipment needed for year-round operation. Because the hauling contractor is hauling year-round, they can a) afford to invest in higher capacity industrial-grade equipment designed for up to 5,000 hour/year (or more) operation, and b) their labor force will develop expertise at the operations, and the Mg handled per unit of equipment investment will be a maximum. These two factors together create the potential to minimize hauling cost ($/Mg).

Storage

The logistics system has three storage features. Round bale packages act as self-storage and protect the biomass. Rounded top sheds water, so the round bale can be left in the field for a short time before in-field hauling. This provides the advantage of uncoupling the harvest and in-field hauling operations, and thus provides an opportunity for improving the cost efficiency of both operations. The farmgate contract holder has the opportunity to bale when the weather is right and haul later.

The second storage features are the SSLs. These locations provide the needed transition between in-field hauling and highway hauling. The SSLs will be located so that the Mg-km parameter for each SSL will be not more than 4 km. This means that, averaged across all Mgs stored at that SSL, each Mg will be hauled less than 4 km from the production field to the SSL. This constraint gives the producer an upper bound for calculating in-field hauling cost and all farmgate contractors are treated the same relative to in-field hauling cost.

The third storage feature is the inventory in at-plant storage, which provides the needed feedstock buffer at the plant. Those building a bioenergy plant would like to operate with just-in-time (JIT) delivery of feedstock as this gives them the lowest cost for receiving facility operation. If JIT is not possible, they want the smallest at-plant inventory for cost effective operation. There is obviously a trade-off in the logistics system design between the higher cost to purchase JIT delivery, and the cost of at-plant storage operations.

At-Plant Storage

In day-to-day management, it will be very difficult to achieve JIT delivery of any feedstock for 24/7 operation. All multi-bale handling unit concepts must include some at-plant storage. Even when known quantities of feedstock are stored in a network of SSLs, and a Feedstock Manager is controlling the deliveries, there will be delays.

To give a frame of reference, suppose 3 days of at-plant storage is the design goal. This inventory amount may suffice in the Southeast where ice and snow on the roads is not typically a significant problem for winter operations. But in the Midwest, additional days of at-plant storage will be required. A visualization of 3-day at-plant storage is shown in Figure 10. The number of racks shown is not part of the "cost analysis" example given later.

Bales will remain in the rack until processed; there is no individual bale handling at the plant. This is a very important aspect of any multi-bale handling system. This reduction in bale handling not only reduces cost, but also reduces damage to the bales. The integrity of the bales will be maintained, and bales will have additional protection from the rain since the rack has a top cover.

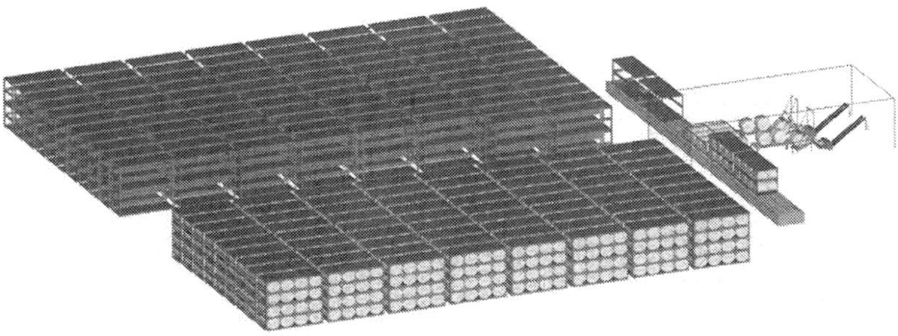

Figure 10: Illustration of at-plant storage for the rack system. At-plant storage shown with 144 racks with bales, 5 racks on conveyor (2 full, 1 being emptied, 2 empty), and 176 empty racks.

Functionality Analysis for Rack System Concept

In this example, the analysis begins with round bales in ambient storage in SSLs and ends with a stream of size-reduced material entering the bioenergy plant for 24/7 operation.

List of Baseline Constraints

The constraints for this example are:

1. The rack design, and the design of the trailer to haul the rack, must conform to the standards for highway transport.
2. The rack used for this example holds sixteen 5×4 round bales. This round bale has a diameter of 1.52-m (5 ft) and a length of 1.22-m (4 ft). With two racks on a truck, a truckload is 32 this kind of round bales.
3. The bales will be pre-loaded into the racks while the trailer is parked at the SSL. When a truck arrives, a trailer with two empty

racks is dropped, and the trailer with two full racks is towed away. The goal for the load time (trailer exchange) is 10 min, or less.

4. Hauling of the trailer and racks will be done 24 hours/day. Truck drivers will work five, 8-hour shifts per week and operations will be continuous for a 6-day work week. The rack loading crew at an SSL will be organized such that each worker will work a 40-h week. SSL operations will load bales into racks six 10-h workdays per week.
5. At the Receiving Facility, the racks are lifted from the trailer and placed on a conveyor to be conveyed into the plant for immediate use, or stacked two high in at-plant storage. The goal at the Receiving Facility is for the forklift to lift two full racks from the trailer and replace them with empty racks within of 10 min, or less.
6. The plant will operate with a maximum of 3 days of at-plant storage.
7. The example assumes that the plant processes one bale per minute. Assuming a bale weighs 900 lbs (0.408 Mg) in average (15% M.C.). The dry mass (DM) per bale is,

$$0.408 \text{ Mg/bale} \times (1-0.15) = 0.3468 \text{ Mg DM/bale} \quad (2)$$

Assuming one bale is processed per minute, the processing rate is,

$$0.33468 \text{ Mg DM/bale} \times 1 \text{ bale/min} \times 60 \text{ min/h} = 20.8 \text{ Mg DM/h} \quad (3)$$

For comparison, 20.8 Mg DM/h is 500 Mg DM/d. This size is in the 100 to 1000 Mg/d range recommended for Regional Biomass Processing Depots by Eranki et al. [54].

The processing time for racks is:

$$60 \text{ bale/h} 16 \text{ bale/rack} = 3.75 \text{ rack/h} \quad (4)$$

Thus, the number of truckloads consumed by the plant for a 24/7 operation is,

$$3.75 \text{ rack/h} \times 24 \text{ h/d} \times 7 \text{ d/wk} = 630 \text{ rack/wk} = 315 \text{ truck loads/wk} \quad (5)$$

Operation Plan for 24-H Hauling

The plant will operate 24/7, but the Receiving Facility will be open 24 hours per day for 6 days. At 6:00 Sunday morning, there will be enough feedstock accumulated in at-plant storage for the maximum 3-day buffer (72 hours). Hauling operations begin again at 6:00 Monday morning. At-plant inventory is decreased to a 2-day buffer (48 hours) to supply the material for operation from 0600 Sunday morning to 6:00 Monday morning.

To discuss the 24-h-hauling concept, it is convenient to consider SSL operations as: 1) "day-haul" operation, and 2) "night-haul" operations. For the day-haul operation, the racks are transported as they are filled during the day. For the night-haul operation, the required number of empty racks, enough for one day's operation, are pre-positioned at the SSL. Cost of pre-positioning the racks is not considered in this example. The SSL crew fills these racks with bales during their 10-h workday and they are hauled during the night. Each truck arriving during the night unhooks a trailer with two empty racks and hooks up a trailer with two full racks. The next morning, the SSL crew will fill the empty racks delivered during the night and fill them during their workday.

Operational Plan for Receiving Facility

A forklift (10-ton or 9.07-Mg capacity) will operate continuously at the receiving facility. This forklift will lift two full racks from the trailer and place them onto a conveyor into the plant for direct processing, or stack these racks in at-plant storage. Then the forklift will lift two empty racks onto the trailer and then the truck returns to the SSL. Empty racks will be stacked in the storage yard until they are lifted onto trailers.

The operational plan calls for two forklifts at the receiving facility, identified as a "work horse" and a "backup." The workhorse will operate continuously and the backup will operate during the day when trucks are waiting in the queue. Key point – the system must have a backup forklift because, if a forklift is not available to lift racks off of and onto trailers, all operations cease.

The handling of the racks emulates the handling of bins at a sugar mill in South Florida. In the bin system, a truck has three bins, two on the first trailer and one on a "pup" trailer. The bins are side-dumped if material is processed directly (Figure 11), or the bins are off-loaded and stacked two-high in the storage yard for nighttime operation (Figure 12). When the bins are dumped directly, it takes 3 min to dump the bins. For normal operation, one truck hauls 10 loads (30 bins) a day. At 37 tons/load (33.6 Mg/load), each truck hauls 370 ton/d (336 Mg/d). Sugar cane is 80% moisture content, so 370 ton = 74 dry ton/d/truck (67.2 Mg DM/d/truck).

Figure 11: Bins being side-dumped at sugar mill in South Florida, USA.

The conveyor for moving the racks into the plant is an adaptation of a piece of commercial technology. Conveyor design and function is similar to the conveyor used to move cotton modules into a cotton gin. Cotton modules and the 16-bale racks have approximately the same dimensions. The conveyor cost used in the analysis was obtained from a cotton equipment manufacturer. The receiving facility operates 6 d/wk, thus, on average, the daily delivery will be:

$$630 \text{ racks/wk} 6 \text{ d/wk} = 105 \text{ racks/d} = 53 \text{ trucks/d} \quad (6)$$

For this example, plant size (23 dry ton/h or 20.9 Mg DM/h) was chosen based on the expected capacity of the two forklifts at the plant. One forklift is expected to lift two full racks and lift and load two empty racks on a trailer at the rate of one truck every 27 minutes averaged over the 24-h a day. The design of the receiving facility and at-plant storage area has to facilitate this operation. A larger at-plant storage area will lower the forklift productivity (ton/h) because the average cycle time to move an individual rack is greater.

Size of At-Plant Storage Yard

For example, maximum at-plant storage for a 3-day supply is, 3.75 racks/h times 72 h, or 270 racks. Racks will be stacked two high in "units" with two rows of 24 spaces each (Figure 12), thus there are 48 storage spaces in each unit. Each unit stores 96 racks. Three units are required for 270-racks storage.

$$3.75 \text{ racks/h} \quad \times \quad 72 \text{ h} = 270 \text{ racks} \tag{7}$$

If 7-day at-plant storage is required, the total number of racks increases to 630 racks, or seven 96-rack units as shown in Figure 14. This implementation of a 7-day supply is not believed to be a cost effective choice, not only because of the capital investment in the racks, but also because of the forklift operating time required to cycle back and forth over a storage area this large. This is a key issue---the larger the at-plant storage, the lower the forklift productivity (ton/h), and thus the higher the forklift cost ($/ton).

The rack system competes best when the racks are filled and emptied as many times as possible in a given time period, not when they act as storage units. Other multi-bale handling units are more suitable for a larger at-plant storage like the one shown in Figure 14.

Figure 12: Bins being stacked in at-plant storage at sugar mill in South Florida, USA.

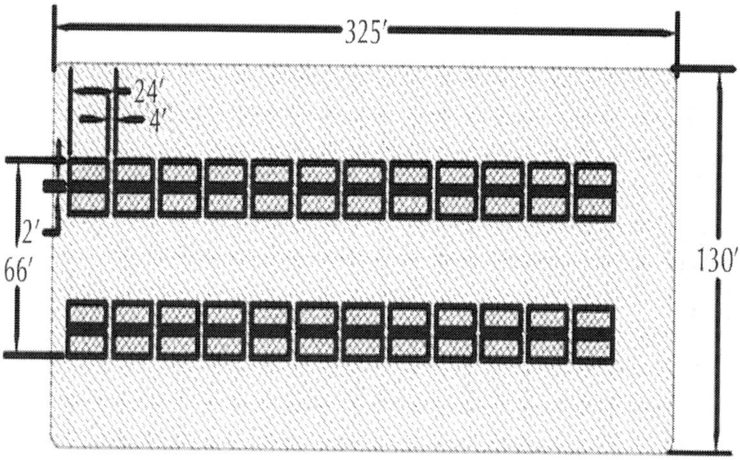

Figure 13: Sample layout of at-plant storage showing a unit (48 storage spaces). Racks are stacked two high for a total capacity of 96 racks (Total area required is approximately 4047 m^2).

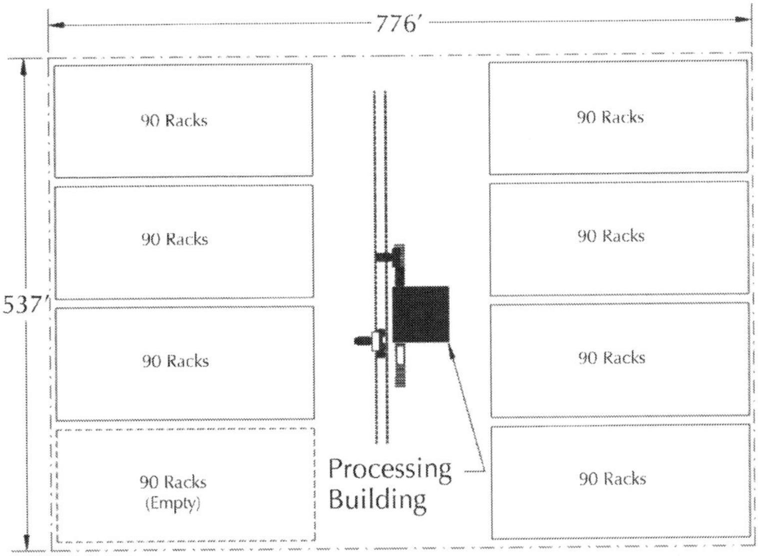

Figure 14: Sample layout of at-plant storage showing a 7-day supply of full racks and space for storing one unit of empty racks. (Total area required is approximately 38445 m²).

Bale Loading Into Racks at the SSL

Ideally, the bale- loading- into- racks operation at the SSL should be able to load 16 bales in a rack in 20 minutes. This design criterion has not been attained with actual equipment.

For a workday with 10 productive hours (600 min), a 20-min-per-rack operation can theoretically load 30 racks, or 15 truckloads. Given the reality of SSL conditions, an actual operation could not sustain this productivity. In this analysis, it is reasonable to assume that a mature operation can average 70% of the theoretical productivity. The number of loads-day-operation used for this analysis will be reduced to 15×0.7 = 10.5 loads/d. The number of individual SSL operations required is 315 loads per week, divided by 6-day SSL operation per week, which is 52.5 or 53 loads per day average operation. This is five operations averaging 10.5 loads per day.

Thus, loading operations will occur at five different SSLs with a different set of equipment for each operation. Each SSL operation will load an average of 21 racks per day. Time to move the crew, equipment and reposition the empty trailer/rack from one SSL to the next is not dealt with in this analysis. Thus, the 21 racks/d productivity may be optimistic.

There are several options to load the rack. The option which found to be the most cost effective is the "side-load" option [58]. A telehandler with special attachment can pick up two bales per cycle (Figure 15) and load these bales into the side of the rack. Assuming that the average productivity to be achieved under production conditions is 34 min per rack, then the time needed to load two racks on a trailer is 68 min. Remember, this is the assumed average load time for year-round operation.

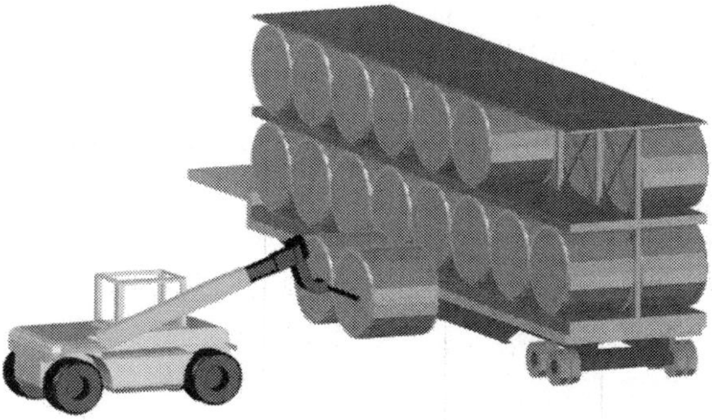

Figure 15: Concept for side loading bales into rack on trailer.

At the SSL, loading of the racks is the most challenging design for a cost-effective biomass logistics system. It is difficult to reduce the cost of these operations because the labor productivity (tons handled per worker-hour) might be low. By uncoupling the SSL loading and highway hauling, the truck does not have to wait for racks to be loaded in order to pick up the trailer. Also, the SSL crew does not have to wait for a truck to arrive with empty racks. However, the hooking/unhooking of trailers may be problematic for some drivers.

The day-haul operations are uncoupled by providing two extra trailers at the "day haul" SSL, and the night-haul operations are uncoupled by providing 9 extra trailers for the "night haul' SSL. Each truck tractor then has 11 extra trailers (22 racks) in the system. This may not be a best (least cost) approach, but it does provide a reasonable starting point.

Influence of SSL Size on Rack Loading Operations

If the yield of switchgrass in a commercial-scale operation averages 4 ton/ac (8.96 Mg/ha), this equates to approximately 9 bales/ac (22 bales/ha). It is suggested that the minimum size of a SSL contain the bales from 60 ac (24.5 ha) is 540 bales, and the maximum size SSL approximately stores bales from 1,200 ac (487 ha) 10,800. If 70% of the theoretical loading (30 racks/d) is achievable, the contactor will load 21 racks times 16 bales/rack is equal to 336 bales per day. The minimum size SSL will remove all of the bales in about 540/336, which is 1.6 days. The maximum size SSL will be loaded out in 10,800/336 = 32 days.

Cost of SSL operations at the smaller SSLs will be higher because of the mobilization charge to move equipment and crew to the next SSL for a fewer days of operation. It is probable that the load-haul contract offered by the biorefinery for each individual SSL will consider the haul distance and SSL size and thus the per-ton price will be different for each SSL.

Total Trucks Required – 24-H Hauling

To achieve 24-h hauling, the truck drivers will work 8-h shifts and the trucks will run continuously from 0600 Monday to 0600 Sunday, a total of 144 h/wk. The total racks processed each week are 630, equal to 315 truckloads. If a uniform delivery is assumed, the average truck unload time is,

$$315 \text{ trucks} / 144 \text{ h} = 2.2 \text{ trucks/h} = 1 \text{ truck/27 min.} \qquad (8)$$

This productivity is well within the Rack System design goal of a 10 min unload time. As previously stated, the 24-h hauling concept envisions that the SSL crew will leave a supply of loaded racks on trailers at the SSL when they finish their 10-h workday. These racks will be hauled during the night. The next morning, the loading crew will load empty racks delivered during the night and load them during their workday.

The key variable in hauling is the truck cycle time. To calculate cycle time, for example, an average haul distance is needed. An actual database was developed for a proposed bioenergy plant location at Gretna, Virginia, USA to calculate the average haul distance.

An analysis was done for a 30-mi (48 km) radius around Gretna to identify potential production fields based on current land use determined from current aerial photography and GIS methods. It is conservatively assumed that 5% of the total land was assigned into switchgrass production. SSLs were established at 199 locations (Figure 5), and the existing road network was used to determine the travel distance from each SSL to the proposed plant location at Gretna. Some loads were hauled 2 miles (3.2 km) and some were hauled over 40 miles (64 km). A weighted ton-mi parameter was computed and found to be 25.4 miles (40.6 km). This means that, across all 199 SSLs, each ton travels an average of 25.4 miles (40.6 km) to get to the plant.

Truck cycle time is calculated using the 25.4-mile average haul distance, a 45 mph average speed, 10-min to hook/unhook trailers a SSL, and 10-min to lift full and empty racks from the trailers. Theoretical cycle time is 1.46 h. In 24 hours of operation, one truck can haul:

$$24 \text{ h}/(1.46 \text{ h/load}) = 16.4 \text{ loads per truck per day} \quad (9)$$

Assuming that a truck fleet can average 70% of the theoretical capacity, then (0.7×16.4 =11.5 loads per truck per day can be achieved. Remember, since the trucks run continuously, a decimal number of loads can be used as the average achieved per-day of productivity.

It is not practical to use the each-haul contractor-runs-their-own-trucks assumption for 24-h hauling. The way to maximize truck fleet productivity is to have the Feedstock Manager have the control to send any truck to any SSL where a trailer with full racks is available. This greatly facilitates the hauling at both day-haul and night-haul SSLs.

Total trucks being controlled by the Feedstock Manager is:

53 loads/day required at the plant/11.5 loads per truck = 4.6 trucks (5 trucks) (10)

Since five SSL crews, one per SSL, are loading trailers/racks. More realistic is that eight to ten trucks will be available and some would also have responsible for bring other supplies to the biorefinery or hauling waste and other value-added products from the biorefinery.

Total Racks and Assets Required For 24-H Hauling

Since the only time deliveries are not being made is the 24-h period from 0600 Sunday to 0600 Monday, the amount in at-plant storage can be reduced. Using 1.5 days as the minimum at-plant storage, so the total capacity hours required in at-plant storage at 0600 Sunday, when deliveries are ended for the week, is

$$24 \text{ h (actual)} + 1.5 \text{ d} \times 24 \text{ h/d (at-plant storage)} = 60 \text{ hours} \qquad (11)$$

$$3.75 \text{ racks/h} \times 60 \text{ h} = 225 \text{ racks} \qquad (12)$$

Total trailers are calculated as follows. Each truck has one trailer connected, two parked at a "day-haul" SSL and nine parked at a "night-haul" SSL for a total of 12 trailers. The total racks on trailers are calculated as:

$$5 \text{ trucks} \times 12 \text{ trailers per truck} \times 2 \text{ racks/trailer} = 120 \text{ racks} \qquad (13)$$

Total racks required are:

At-plant	+	On 60 trailers	+	Reserve	=	Total
225	+	120	+	5	=	350

Loaded trailers can be parked at the receiving facility when hauling ends for the week at 0600 Sunday. In order for this procedure to work, the racks on most of these 60 trailers have to be returned to SSLs during the period 0600 Sunday to 0600 Monday so they will be in position for operations to begin at each SSL at 0600 Monday. This requires some empty back hauls. Cost for these empty back hauls is a level of detail that must wait for a more sophisticated analysis.

When racks on trailers are counted as part of the at-plant storage, the minimum number of racks is: racks processed per year divided by 230, that is 129 cycles per year; or about 2.7 cycles per week for 47 weeks of annual operation.

At-plant	+	On 60 trailers	+	Reserve	=	Total
(225-120)	+	120	+	5	=	230

Cost Analysis for 24-H Hauling Using Rack System

The costs given in this section are presented without supporting detail. They were calculated using the procedures given in [69]. They are "best estimates" given current cost parameters. All costs are given on a \$/Mg DM basis for operation of a bioenergy plant consuming 23 dry ton/h (20.9 Mg DM/h). The challenge is to find a way that machine productivity (Mg/h) can be increased.

Total Truck Cost

The assumed truck cost (tractor and trailer for hauling the two racks) is \$630/d for a 24-h workday, which includes ownership plus operating cost, plus labor, but excluding fuel.

Truck cost, excluding fuel, is:

$$\$630/\text{d}11.5\ \text{loads/d} \times 12.2\ \text{dry ton/load} = \$\ 4.49/\text{dry ton}=\$4.95/\text{Mg DM} \quad (14)$$

Truck fuel cost for the 25.4 miles (40.6 km) average haul distance is:

$$(25.4\ \text{mi} \times 2)/4\ \text{mi/gal} = 12.7\ \text{gal} \times \$3.50/\text{gal} = \$44.45/\text{load} \quad (15)$$

$44.45/load/ 12.2 dry ton/load = $3.64/dry ton ($4.01/Mg DM) (16)

Total truck cost is ownership and operating plus fuel, that is 4.95 plus 4.01 = $8.96/Mg DM.

Load, Unload Operations

1. Handling racks at plant: $1.93 (forklift) + $1.02 (backup forklift) = $2.95/dry ton
2. SSL *operation*: $3.66 (telehandler) + $0.98 (extra trailers) = $4.64/dry ton
3. Rack cost: cost for 230 racks = $1.80/dry ton
4. Storage yard at processing plant: $0.13/dry ton
5. Conveyor entering plant: $0.28/dry ton

(Note: 1 dry ton = 0.907 Mg DM)

Size Reduction

1. Unroller-chopper cost: $5.76/dry ton.
2. Rack cost: all costs associated with the ownership and maintenance of the racks.
3. Loading cost: all costs associated with the loading of bales into racks. These costs are referred to as "SSL operation costs".
4. Truck cost: all costs associated with the ownership and operation of the trucks.
5. Receiving Facility cost: all costs associated with the unloading of racks from trucks, placement of racks onto conveyor (or placement in at-plant storage), conveyor operation, operation of at-plant storage, and removal of racks from at-plant storage and placement on trucks for return to SSL.
6. Size reduction: all costs associated with the unloading of bales from the rack, operation of conveyor for single file bales delivered to size reduction machine, and operation of machine for initial size reduction.

Truck cost is 34% of the total cost, SSL operations are 20%, Receiving Facility operations are 14%, size reduction is 24%, and the racks are 8%. It is clear why the Rack System Concept was organized to maximize truck productivity; truck cost is the largest cost component. Truck cost plus SSL operations are $12.77/dry ton, or 54% of total cost. The Receiving Facility cost is $3.37/dry ton, only 14% of total cost. As with all other multi-bale handling system concepts, the Rack System provides an opportunity for minimizing cost between the plant gate and the size reduction unit operation.

The total cost shown in Table 2 does not include the farmgate contract cost (production, harvesting, in-field transport, storage in SSL, and profit to producer). The farmgate contract cost can be estimated from local data for production, harvest, and ambient storage of round bales of hay. In the Southeast the key issue relative to the hay cost comparison is the difference in yield; switchgrass will yield about 9 Mg/ha as compared to traditional hay species that yield about 4.5 Mg/ha.

Table 2: Total cost for hauling, receiving facility operations, and size reduction for rack system example – side-load option, 24-h hauling.* $1/dry ton = $1.103/Mg DM

Operation	**Cost ($/dry ton)***
Racks	1.80
Loading at SSL	
Telehandler	3.66
Extra Drop-deck Trailers	0.98
Truck cost	8.13
Unloading at plant	
Workhorse forklift	1.93
Backup forklift	1.02
At-plant storage (Gravel lot with lighting)	0.13
Conveyor into plant	0.28
Unroller-chopper (Initial size reduction)	5.76
Total	$23.69

CONCLUSIONS

The key decision points for the design of a logistics system for a bioenergy plant operating 24/7 year-round are summarized as follows.

1. A complete logistics system is defined as one that begins with the biomass standing in the field and ends with a stream of size-reduced material entering a bioenergy plant for 24/7 operation. Optimizing one unit operation in isolation may increase the cost of an "upstream" or "downstream" operation such that total delivered cost is increased.
2. Most feedstock is harvested only part of the year, thus storage is a part of the logistics system. A cost effective logistics system provides for efficient flow of material in and out of storage.
3. Just-in-time (JIT) delivery of feedstock provides for a minimum at-plant storage cost. Since JIT delivery is not practical for typical biomass logistics systems, there is always a cost trade-off between the size of at-plant storage and the other design constraints needed to insure a continuous feedstock supply. Knowledge of quantities of biomass in Satellite Storage Locations (SSLs) provides the Feedstock Manager at a bioenergy plant an opportunity to minimize the at-plant storage cost.
4. Farmgate contracts that require a summer-early fall harvest must compensate for the removal of nutrients, and contracts that require a winter harvest must compensate for loss of yield incurred by the delayed harvest.
5. Assigning different unit operations to different entities in the business plan can lower average delivered cost. For example, it is more efficient to pool all farmgate activities into a farmgate contract and have a hauling contractor handle all load-haul activities. This division is defined as a division between "agricultural" and "industrial" operations. The key benefit achieved is in the capitalization of the equipment. Load-haul contractors can afford to invest in industrial-grade, high-capacity equipment designed for year-round operation as compared to farmgate contractors who will use their equipment 400 hours (or less) per year.
6. Uncoupling of the unit operations in the logistics chain can provide an advantage.

7. In the agricultural operations, baling uncouples the harvesting and in-field hauling operations. When the harvesting operation is not constrained by in-field hauling---both unit operations can proceed at maximum productivity.
8. In the industrial operations, it is important to uncouple truck loading from hauling. Maximum loads-per-day-per-truck are achieved when the loading crew never has to wait for a truck to arrive and the truck never has to wait to be loaded.
9. Truck cost is the largest component of total cost in most logistics systems, thus it is essential to maximize truck productivity (Mg hauled per unit time) by increasing both Mg-per-load and loads-per-day. A 10-min load time and a 10-min unload time is a desired goal for design of most logistic systems.
10. Multi-bale handling units are in need to solve the rapid loading/ unloading challenge.
11. Twenty-four-hour hauling can minimize truck cost ($/Mg). The challenge is to design a logistics system with a practical procedure for loading trucks at night at a remote location.
12. The design of the Receiving Facility, because of the need to unload trucks quickly, is critical in the design of a complete logistics system. Typically, this design specifies that each load have the same configuration, and requires a delivery schedule where approximately the same number of loads is received each workday.
13. The most cost-effective logistics system will be structured such that information technologies (GPS, bar codes, entry of data over cell phone network) and optimization routines developed for other logistics systems can be used to optimize asset utilization in real time.

REFERENCES

1. D K. Brownell, J. Liu, J W. Hilton, T L. Richard, G R. Cauffman, and MacAfee B R (2009Evaluation of two forage harvesting systems for herbaceous biomass harvestingASABE Paper 097390St. Joseph, MI: ASABE.

2. S G. Reynolds, and J. Frame, 2005Grasslands: Developments, opportunities, perspectives.Enfield, New Hampshire: Science Publ. Inc.
3. J S. Cundiff, and R D. Grisso, 2008Containerized handling to minimize hauling cost of herbaceous biomassBiomass & Bioenergy 324308313
4. E V. Hierden, 1999Apparatus for processing large square hay bales into smaller recompressed Bales. U.S. Patent 6339986
5. Steffen Systems (2009) High Density Bale Compression Systems. Available at: http://www.steffensystems.com/Products/Bale_Press/index.htm.Accessed 2012 April 19.
6. S. Sokhansanj, A F. Turholow, J. Stephen, M. Stumborg, J. Fenton, and S. Mani, 2008Analysis of five simulated straw harvest scenarios. Can. Biosys. Eng. 50:2.272
7. S. Sokhansanj, A F. Turhollow, and E G. Wilkerson, 2008Development of the Integrated Biomass Supply Analysis and Logistics Model (IBSAL)Technical Memorandum ORNL/TM-2006/57. Oak Ridge, TN: Oak Ridge National Laboratory.
8. P. K. Kumar, and K. E. Ileleji, 2009Tech-no-economic analysis of the transportation, storage, and handling requirements for supplying lignocellulosic biomass feedstocks for ethanol production. ASABE Paper 097427St. Joseph, MI: ASABE.
9. D K. Brownell, 2010Analysis of biomass harvest, handling, and computer modelingM.S. Thesis. The Pennsylvania State University, University Park, PA.
10. A K. Srivastava, E G. Carroll, R P. Rohrbach, and D R. Buchmaster, 2006Engineering principles of Agricultural MachinesAmerican Society of Agricultural and Biological Engineers, St. Joseph, MI: ASABE.
11. J S. Cundiff, R D. Grisso, and D. Mccullough, 2011Comparison of bale operations for smaller production fields in the Southeast. ASABE Paper 1110922St. Joseph, MI: ASABE.
12. ASAE Standards D4972006Agricultural machinery management data. American Society of Agricultural Engineers, St. Joseph, MI: ASAE.
13. J Liu, and B Kemmerer, (2011) Field performance analysis of a tractor and a large square baler. SAE Technical Paper 2011-01-

2302, Presented at Commercial Vehicle Engineering Congress, Chicago, IL doi:10.4271/2011-01-2302.

14. B. Kemmerer, and J. Liu, 2012Large square baling and bale handling efficiency-A case study. Ag. Sci. 32178183
15. ASAE Standards EP4962006Agricultural machinery management. American Society of Agricultural Engineers, St. Joseph, MI: ASABE.
16. J S. Cundiff, J H. Fike, D J. Parrish, and J. Alwang, 2009Logistic constraints in developing dedicated large-scale bioenergy systems in the Southeastern United StatesJournal of Environmental EngineeringASCE 1351110861096
17. G. Hadders, and R. Olsson, 1997Harvest of grass for combustion in late summer and in springBiomass Bioenergy 123171175
18. J A. Larson, D F. Mooney, B C. English, and D D. Tyler, 2010Cost analysis of alternative harvest and storage methods for switchgrass in the Southeastern U.S. 2010 Annual Meeting of Southern Agricultural Economics Association, Feb. 6-9, 2010, Orlando, FL.
19. F M. Epplin, L. Mapemba, and G. Tembo, 2005Economic modeling of a lignocellulosic biomass biorefining industry. In: Outlaw JL, Collins KJ, Duffield JA, editors. Agriculture as a producer and consumer of energy. London: CABI Publishing; 2005. 205217
20. S. Hwang, F M. Epplin, B. Lee, and R. Huhnke, 2009A probabilistic estimate of the frequency of mowing and baling days available in Oklahoma USA for the harvest of switchgrass for use in biorefineriesBiomass Bioenergy 33810371045
21. M A. Sanderson, R P. Egg, and A E. Wiselogel, 1997Biomass losses during harvest and storage of switchgrassBiomass Bioenergy 122107114
22. A. Kumar, and S. Sokhansanj, 2007Switchgrass (Panicum vigratum,L.) delivery to a biorefinery using integrated biomass supply analysis and logistics (IBSAL) model. Bioresource Technology 9810331044
23. D. Rees, V H (1982A discussion of the sources of dry matter loss during the process of haymaking. J. of Agric. Engng Res. 27469479

24. C G. Coble, and R. Egg, 1987Dry matter losses during hay production and storage of sweet sorghum used for methane productionBiomass 14209217
25. B L. Bledsoe, and B M. Bales, 1992Bale wrap evaluation. ASAE Paper 921573St. Joseph, MI: ASAE.
26. LaFlamme L F (1989Effects of storage conditions for large round bales on quality of grass-legume hayCan. J. of Animal Sci. 69955961
27. A. Turhollow, 1994The economics of energy crop productionBiomass Bioenergy 63229241
28. K J. Shinners, G C. Boettcher, R E. Muck, P J. Wiemer, and M D. Casler, 2010Harvest and storage of two perennial grasses as biomass feedstocks. Trans. of ASABE 532359370
29. Chariton Valley RC&D (2002Chariton Valley Biomass Project - Draft Fuel Supply PlanUnited States Department of Energy, Contract Number: DE-FC36-96GO10148
30. D. Styles, F. Thorne, and M B. Jones, 2008Energy crops in Ireland: an economic comparison of willow and Miscanthus production with conventional farming systemsBiomass Bioenergy 325407421
31. M. Khanna, B. Dhungana, and J. Clifton-brown, 2008Costs of producing Miscanthus and switchgrass for bioenergy in IllinoisBiomass Bioenergy 326482493
32. A. Monti, S. Fazio, V. Lychnaras, P. Soldatos, and G. Venturi, 2007A full economic analysis of switchgrass under different scenarios in Italy estimated by BEE modelBiomass Bioenergy 314177185
33. Duffy M (2007) Estimated costs for production, storage and transportation of switchgrass. Ames, IA: Iowa State University, Extension Report No. PM 2042
34. R M. De Mol, M. Jogems, A H. Van Beek, P. Gigler, J K. Simulation, and optimization of the logistics of biomass fuel collection. Neth J Agric Sci 45219228
35. J. Cundiff, N. Dias, and H D. Sherali, 1997A linear programming approach for designing a herbaceous biomass delivery systemBioresource Technology594755

36. M. Zuo, W. Kuo, and K L. Mcroberts, 1991Application of mathematical programming to a large-scale agricultural production and distribution systemJ Oper Res Soc 428639648
37. L D. Mapemba, F M. Epplin, R L. Huhnke, and C M. Taliaferro, 2008Herbaceous plant biomass harvest and delivery cost with harvest segmented by month and number of harvest machines endogenously determinedBiomass Bioenergy 321110161027
38. A. Kumar, S. Sokhansanj, and P C. Flynn, 2006Development of a multicriteria assessment model for ranking biomass feedstock collection and transportation systemsAppl Biochem Biotechnol 129-132:71 EOF87 EOF
39. A. Mukunda, K E. Ileleji, and H. Wan, 2006Simulation of corn stover logistics from on-farm storage to an ethanol plantASABE Paper 066177St. Joseph, MI: ASABE.
40. P P. Ravula, R D. Grisso, and J S. Cundiff, 2008Cotton logistics as a model for a biomass transportation systemBiomass & Bioenergy 324314325
41. S. Sokhansanj, A. Kumar, and A F. Turhollow, 2006Development and implementation of integrated biomass supply analysis and logistics model (IBSAL)Biomass Bioenergy 30838847
42. Y. Shastri, A. Hansen, L. Rodriguez, and K C. Ting, 2011Optimization of Miscanthus harvesting and handling as an energy crop: BioFeed model applicationBiol Eng 313769
43. Y.Shastri,A.Hansen,L.Rodriguez,andKC.Ting,2011Development and application of BioFeed model for optimization of herbaceous biomass feedstock productionBiomass Bioenergy 3529612974
44. C T. Gallis, 1996Activity oriented stochastic computer simulation of forest biomass logistics in GreeceBiomass Bioenergy 10377382
45. D. Nilsson, 2000Dynamic simulation of straw harvesting systems: influence of climatic, geographical and biological factors on performance and costsJ. Agric. Engng Res 762736
46. D. Freppaz, R. Minciardi, M. Robba, M. Rovatti, R. Sacile, and A. Taramasso, 2004Optimizing forest biomass exploitation for energy supply at a regional levelBiomass Bioenergy 261525
47. J D. Judd, S C. Sarin, J S. Cundiff, and R D. Grisso, 2010Cost analysis of a biomass logistics system. ASABE Paper 1110466St. Joseph, MI: ASABE.

48. I. Tatsiopoulos, and A. Tolis, 2003Economic aspects of the cotton-stalk biomass logistics and comparison of supply chain methodsBiomass Bioenergy 24199214
49. G. Tembo, F M. Epplin, and R L. Huhnke, 2003Integrative investment appraisal of a lignocellulosic biomass-to-ethanol industryJ. Agric. Res. Econ. 28611633
50. V R. Morey, N. Kaliyan, D G. Tiffany, and D R. Schmidt, 2010A corn stover supply logistics system. Applied Eng. in Agric. 26(3):455-461.
51. J. Nagel, 2000Determination of an economic energy supply structure based on biomass using a mixed-integer linear optimization modelEcol. Engng 1691102
52. J. Gan, and C T. Smith, 2011Optimal plant size and feedstock supply radius: A modeling approach to minimize bioenergy production costsBiomass Bioenergy 35833503359
53. A. Dunnett, C. Agjiman, and N. Shah, 2007Biomass to heat supply chains applications of process optimizationTransactions Institution of Chemical Engineers: Process Safety and Environmental ProtectionB5): 419 EOF429 EOF
54. M. Bruglieri, and L. Liberti, 2008Optimal running and planning of a biomass-based energy production processEnergy Policy36724302438
55. S. Leduc, D. Schwab, E. Dotzauer, E. Schmid, and M. Obersteiner, 2008Optimal location of wood gasification plants for methanol production with heat recoveryInt'l J. of Energy Res. 321210801091
56. R. Berruto, and P. Busato, 2008System approach to biomass harvest operations: simulation modeling and linear programming for logistic designASABE Paper 084565St. Joseph, MI: ASABE.
57. J P. Resop, J S. Cundiff, and C D. Heatwole, 2010Spatial analysis to site satellite storage locations for herbaceous biomass in the Piedmont of the Southeast. Applied Eng. Agric. 27(1): 25-32.
58. J D. Judd, S C. Sarin, and J S. Cundiff, 2011An optimal storage and transportation system for a cellulosic ethanol bio-energy plantASABE Paper 109413St. Joseph, MI: ASABE.
59. J S. Cundiff, R D. Grisso, and P P. Ravula, 2004Management system for biomass delivery at a conversion plant. ASAE/CSAE Paper 046169St. Joseph, MI: ASABE

60. P.L Eranki, ., D Bryan, . Bals, and Bruce E. Dale. 2011. Advanced regional biomass processing depots: a key to the logistical challenges of the cellulosic biofuel industry. Biofuels, Bioprod. Bioref. DOI:10.1002/bbb. Society of Chemical Industry and John Wiley & Sons, Ltd.

61. T. Yoshioka, K. Aruga, T. Nitami, H. Sakai, and H. Kobayashi, 2006A case study on the costs and the fuel consumption of harvesting, transporting, and chipping chains for logging residues in JapanBiomass Bioenergy 30342348

62. T. Ranta, 2005Logging residues from regeneration fellings for biofuel production-a GIS-based availability analysis in FinlandBiomass Bioenergy 28171182

63. T. Ranta, and S. Rinne, 2006The profitability of transporting uncomminuted raw materials in FinlandBiomass Bioenergy 30231237

64. N. Ayoub, R. Martins, K. Wang, H. Seki, and Y. Naka, 2007Two levels decision system for efficient planning and implementation of bioenergy productionEnergy Conv. Man. 48709723

65. J S. Cundiff, H. Shapouri, and R D. Grisso, 2007Economic analysis of two receiving facility designs for a bioenergy plantASABE Paper 076051St. Joseph, MI: ASABE

66. D. Brownell, and J. Liu, 2012Managing biomass feedstocks: selection of satellite storage locations for different harvesting systemsAgric. Engng Int'l CIGR Journal, 14(1): Manuscript 1886pp.

67. K. Cromer, 2011Personal communication. FDC Enterprises-led Consortium. Antares Group Inc., 57 S. Main St., Suite 506, Harrisonburg, VA 22801. Cooper, Sam, courtesy of Sugar Journal, P.O. Box 19084, New Orleans, LA 70179).

68. Kelderman Mfgto implement multi-bale handling unit.

Citations

CHAPTER 1

Nicolas Carels (2011). The Challenge of Bioenergies: An Overview, Biofuel's Engineering Process Technology, Dr. Marco Aurelio Dos Santos Bernardes (Ed.), ISBN: 978-953-307-480-1.

CHAPTER 2

Tail Lope, Adapt Phani, and Kashaninejad Mahdi, Biomass Feedstock Pre-Processing– Part 1: Pre-Treatment, DOI: 10.5772/17086.

CHAPTER 3

Lope Tabil, Phani Adapa and Mahdi Kashaninejad (2011). Biomass Feedstock Pre-Processing – Part 2: Densification, Biofuel's Engineering

Process Technology, Dr. Marco Aurelio Dos Santos Bernardes (Ed.), ISBN: 978-953-307-480-1, InTech, DOI: 10.5772/18495.

CHAPTER 4

Stéphane C. Marie-Rose, Alexis Lemieux Perinet, and Jean-Michel Lavoie (2011). Conversion of Non-Homogeneous Biomass to Ultraclean Syngas and Catalytic Conversion to Ethanol, Biofuel's Engineering Process Technology, Dr. Marco Aurelio Dos Santos Bernardes (Ed.), ISBN: 978-953-307-480-1, InTech, DOI: 10.5772/20032.

CHAPTER 5

Jernej Mele (2011). Scale-up of a Cold Flow Model of FICFB Biomass Gasification Process to an Industrial Pilot Plant – Example of Dynamic Similarity, Progress in Biomass and Bioenergy Production, Dr. Shahid Shaukat (Ed.), ISBN: 978-953-307-491-7, InTech, DOI: 10.5772/16439.

CHAPTER 6

Babak Fakhim and Bijan Farhanieh (2011). Second Law Analysis of Bubbling Fluidized Bed Gasifier for Biomass Gasification, Progress in Biomass and Bioenergy Production, Dr. Shahid Shaukat (Ed.), ISBN: 978-953-307-491-7, InTech, DOI: 10.5772/16912.

CHAPTER 7

Jude Liu, Robert Grisso and John Cundiff (2013). Harvest Systems and Analysis for Herbaceous Biomass, Biomass Now - Cultivation and Utilization, Dr. Miodrag Darko Matovic (Ed.), ISBN: 978-953-51-1106-1, InTech, DOI: 10.5772/53875.

Index

D

E

F

G

H

I

K

L

M

N

O

P

R

S